MULTISENSOR
INSTRUMENTATION
6σ DESIGN

MULTISENSOR INSTRUMENTATION 6σ DESIGN

Defined Accuracy Computer-Integrated Measurement Systems

PATRICK H. GARRETT

A Wiley-Interscience Publication
JOHN WILEY & SONS, INC.

Library of Congress Cataloging-in-Publication Data:

Garrett, Patrick H.
    Multisensor instrumentation 6[sigma] design / Patrick H. Garrett
        p. cm.
    Title has numeral 6 followed by Greek sigma.
    "A Wiley-Interscience publication."
    ISBN 0-471-20506-0 (cloth)
        1. Electrooptical devices—Testing—Congresses.   2. Automatic checkout equipment—Congresses.   I. Title.

    TA1750 .G37 2001
    670.42′7—dc21   2001046730

Printed in the United States of America

10 9 8 7 6 5 4 3 2 1

CONTENTS

PREFACE

Over the past decade, technical innovation and evolution in computer-centered measurement systems have led to significant performance advances and economies of scale from manufacturing processes to biotechnology laboratories that share multisensor information system common denominators. This book provides a definitive instrumentation circuit and system reference, supported by 46 tables of engineering data, that demonstrates a composite, error-modeled design methodology for implementing defined accuracy, computer integrated measurement systems. The comprehensive accountability presented is consistent with six-sigma quality metrics, and features a user-interactive analysis spreadsheet for the performance optimization of instrumentation designs through total error minimization.

An expanding reliance on six-sigma methods for process optimization (define-measure-analyze-improve-control) emphasizes the significance of data measurement accountability, especially as it affects integrated information structures and automatic control systems. Accordingly, highlights of this book include end-to-end instrumentation performance modeling in terms of comprehensive mean and RSS errors, signal conditioning capabilities extending to sensor noise thresholds, sampled-data design guided by definitive intersample error metrics, and evaluation of algorithmic error propagation in complex multisensor architectures.

The first chapter presents a compendium of sensors with a signal model hierarchy of ascending complexity from apparatus, to in situ, to analytical measurements. The next three chapters cover linear signal conditioning devices and circuits with five categories of instrumentation amplifier sensor interfaces for the upgrading of microvolt signals immersed in volts of random and coherent interference. In the following two chapters, data conversion devices and their performance are analyzed, including seven application-specific A/D converter types. Digital data conversion system design employs intersample error to evaluate sampled data influences ranging from noise aliasing, to oversampling, to the effectiveness of various output signal interpolation functions, including closed-loop bandwidth in digital control systems.

The final three chapters present diverse instrumentation system examples illustrating this error-modeled design approach from process controllers, to video digitization, to vibration analyzers. Multisensor error propagation is described by detailed examples of sequential, homogeneous, and heterogeneous architectures shown, respectively, by turbine engine airflow, electric machine health monitoring, and materials manufacturing process instrumentation. Instrumentation system integration is then examined in a progression from discrete instruments, to remote I/O, to virtual instruments, to analytical instruments exemplified, respectively, by automatic test equipment, satellite meteorology instrumentation, programmable microwave microscopy, and analytical instruments for advanced control. This taxonomy includes interfaces from FireWire to Gigabit Ethernet. In fact, analytical instrumentation in contemporary process systems typically performs at a higher level of abstraction than associated control algorithms.

Many of the developments presented have not appeared in other books or papers. Notable are derivations of device errors such as filter mean error and signal error quantitation resulting from linear input signal conditioning operations and sampled output signal interpolator effectiveness. The author accepts responsibility for the ideas presented and any shortcomings, and hopes that they may stimulate further study and contribution to these topics.

PATRICK H. GARRETT

1

PROCESS, QUANTUM, AND ANALYTICAL SENSORS

1-0 INTRODUCTION

Automatic test systems, manufacturing process control, analytical instrumentation, and aerospace electronic systems all would have diminished capabilities without the availability of contemporary computer integrated data systems with multisensor information structures. This text develops supporting quantitative error models that enable a unified performance evaluation for the design and analysis of linear and digital instrumentation systems with the goal of compatibility of integration with other enterprise quality representations.

   This chapter specifically describes the front-end electrical sensor devices for a broad range of applications from industrial processes to scientific measurements. Examples include environmental sensors for temperature, pressure, level, and flow; in situ sensors for measurements beyond apparatus boundaries, including spectrometers for chemical analysis; and ex situ analytical sensors for manufactured material and biomedical assays such as microwave microscopy. Hyperspectral sensing of both spatial and spectral data is also introduced for improved understanding through feature characterization. It is notable that owing to advancements in higher attribution sensors, they are increasingly being substituted for process models in many applications.

1-1 INSTRUMENTATION ERROR REPRESENTATION

In this text, error models are derived employing electronic device, circuit, and system parameter values that are combined into a unified end-to-end performance representation for computer-based measurement and control instrumentation. This methodology enables system integration beneficial to contemporary technologies ranging from micromachines to distributed processes. Since the baseline performance of machines and processes can be described by their internal errors, it is axiomatic that their performance may also be optimized through design in pursuit of

1

error minimization. Instrumentation system errors are interpreted graphically in Figure 1-1. Total error is shown as the composite of barred mean error contributions plus the root-sum-square (RSS) of systematic and random uncertainties; the true value is ultimately traceable to a reference calibration standard harbored by NIST. Although total error may instantaneously be greater or less than mean error from the additivity of RSS uncertainty error, throughout this text total error is expressed as the sum of mean and RSS errors in providing accountability of system behavior.

Total error is analytically expressed by equation (1-1) as 0–100% of full scale (%FS), where the RSS sum of variances represents a one-sigma confidence interval. Consequently, total error may be expressed over any confidence interval by adding one mean error value and a summation of RSS error values corresponding to the standard deviation integer. Confidence to six sigma is therefore defined by mean error plus six times the RSS error value. Mean error frequently arises in instrumentation systems from transfer function nonlinearities that, unlike RSS uncertainty error, which may be reduced by averaging identical systems as shown in Chapter 4, Section 4-4, instead increases with the addition of each mean error contribution, necessitating remedy through minimal inclusion. Accuracy is defined as the complement of error (100%FS – ε%FS), where 1%FS error corresponds to 99%FS accuracy.

A six-sigma framework is accordingly offered in terms of models defining multisensor instrumentation errors to provide a generic design and analysis methodology compatible with corollary enterprise-quality representations. Quantitative instrumentation system performance expressed in terms of modeled errors assumes that external calibration is maintained to known standards, as shown in Figure 1-21, verifying zero and full-scale values for external instrumentation registration. Calibration is essential and can be performed manually or by automated methods, but it cannot characterize instrumentation device, circuit, and system variabilities that these error budgets ably describe, including expression to 6σ confidence.

$$\varepsilon_{\text{total}} = \Sigma \; \overline{\varepsilon_{\text{mean}}} \; \%FS + [\Sigma \; \varepsilon_{\text{systematic}}^2 + \Sigma \; \varepsilon_{\text{random}}^2]^{1/2} \; \%FS1\sigma \qquad (1\text{-}1)$$

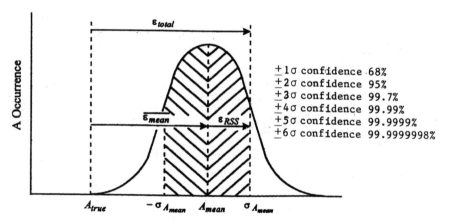

FIGURE 1-1. Instrumentation error interpretation.

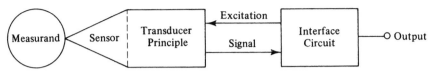

FIGURE 1-2. Generic sensor elements.

Figure 1-2 describes generic measurement elements, where the sensor represents a physical device employed at a measurement interface, and the transducer principle the translation involved between measurand units and a corresponding signal representation. For example, in the application of a thermocouple, the physical contact of two dissimilar alloys with a thermal process constitutes the sensor, but the emf signal arising at discrete temperature values is attributable to the Seebeck transducer effect. Many sensors, such as strain-gauge bridges, further require external excitation to generate a measurement signal as well as a specific interface circuit. Seven measurement definitions follow:

Accuracy: the closeness with which a measurement approaches the true value of a measurand, usually expressed as a percent of full scale

Error: the deviation of a measurement from the true value of a measurand, usually expressed as a percent of full scale

Tolerance: allowable error deviation about a reference of interest

Precision: an expression of a measurement over some span described by the number of significant figures available

Resolution: an expression of the smallest quantity to which a quantity can be represented

Span: an expression of the extent of a measurement between any two limits

Range: an expression of the total extent of measurement values

Technology has advanced significantly as a consequence of sensor development. However, measurement is an inexact discipline requiring the use of reference standards to provide accountability for sensor signals with respect to their measurands. Fortunately, sensor nonlinearity can be minimized by means of multipoint calibration. Practical implementation often requires the synthesis of a linearized output function that achieves the best asymptotic approximation to the true value over a sensor measurement range of interest. The resulting straight-line fit is often realized after digital signal conversion to benefit from the accuracy of digital computation.

The cubic function of equation (1-2) is an effective linearizing equation demonstrated over the full 700°C range of a commonly applied Type-J thermocouple, which is tabulated in Table 1-1. Solution of the A and B coefficients at judiciously spaced temperature values defines the linearizing equation with a 0°C intercept. Evaluation at linearized 100°C intervals throughout the thermocouple range reveals

TABLE 1-1. Sensor Cubic Linearization

$Y \, °C$	$X \, mV$	$y \, °C$	$\varepsilon_{\%FS} = |(Y-y)100\%/700°C|$
0	0	0	0
100	5.269	98	0.27
200	10.779	200	0
300	16.327	302	0.25
400	21.848	401	0.23
500	27.393	500	0
600	33.102	599	0.17
700	39.132	700	0

Y	true temperature	0.11%FS	mean error
X	Type-J thermocouple signal	0°C	intercept
y	linearized temperature	700°C	full scale

temperature values nominally within 1°C of their true temperatures, which correspond to typical errors of 0.25%FS. It is also useful to express the average of discrete errors over the sensor range, obtaining a mean error value of $\overline{0.11}$%FS for the Type-J thermocouple. This example illustrates a design goal proffered throughout this text of not exceeding one-tenth percent error for any contributing system element. Extended polynomials permit further reduction in linearized sensor error while incurring increased computational burden, where a fifth-order equation can beneficially provide linearization to 0.1°C, corresponding to $\overline{0.01}$%FS mean error.

$$y = AX + BX^3 + \text{intercept} \qquad (1\text{-}2)$$

Coefficient for 10.779 mV at 200°C:

$$200°C = A(10.779 \, mV) + B(10.779mV)^3 + 0°C$$

$$A = 18.5546 \frac{°C}{mV} - B(116.186mV^2)$$

Coefficient for 27.393 mV at 500°C:

$$500°C = 508.2662 \, °C - B(3182.68 \, mV^3) + B(20,555.0 \, mV^3)$$

$$A = 18.6099 \qquad B = -0.000475 \frac{°C}{mV^3}$$

1-2   TEMPERATURE SENSORS

Thermocouples are widely used as temperature sensors because of their ruggedness and broad temperature range. Two dissimilar metals are used in the Seebeck effect temperature-to-emf junction with transfer relationships described by Figure 1-3. Proper operation requires the use of a thermocouple reference junction in series with the measurement junction to polarize the direction of current flow and maximize the measurement emf. Omission of the reference junction introduces an uncertainty evident as a lack of measurement repeatability equal to the ambient temperature.

An electronic reference junction that does not require an isolated supply can be realized with an Analog Devices AD590 temperature sensor, as shown in Figure 4-5. This reference junction usually is attached to an input terminal barrier strip in order to track the thermocouple-to-copper circuit connection thermally. The error signal is referenced to the Seebeck coefficients in mV/°C (see Table 1-2) and provided as a compensation signal for ambient temperature variation. The single calibration trim at ambient temperature provides temperature tracking within a few tenths of a °C.

Resistance thermometer devices (RTDs) provide greater resolution and repeatability than thermocouples, the latter typically being limited to approximately 1 °C. RTDs operate on the principle of resistance change as a function of temperature, and are represented by a number of devices. The platinum resistance thermometer is frequently utilized in industrial applications because it offers good accuracy with mechanical and electrical stability. Thermistors are fabricated by sintering a mixture of metal alloys to form a ceramic that exhibits a significant negative temperature coefficient. Metal film resistors have an extended and more linear range than thermistors, but thermistors exhibit approximately ten times their sensitivity. RTDs require excitation, usually provided as a constant current source, in order to convert

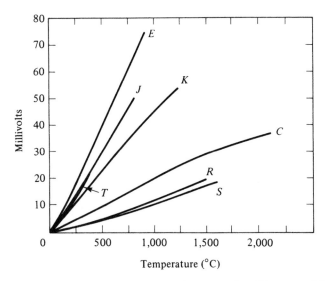

FIGURE 1-3. Temperature–millivolt graph for thermocouples. (Courtesy of Omega Engineering, Inc., an Omega Group Company.)

TABLE 1-2. Thermocouple Comparison Data

Type	Elements, +/–	mV/°C	Range (°C)	Application
E	Chromel/constantan	0.063	0 to 800	High output
J	Iron/constantan	0.054	0 to 700	Reducing atmospheres
K	Chromel/alumel	0.040	0 to 1,200	Oxidizing atmospheres
R&S	Pt-Rb/platinum	0.010	0 to 1,400	Corrosive atmospheres
T	Copper/constantan	0.040	−250 to 350	Moist atmospheres
C	Tungsten/rhenium	0.012	0 to 2,000	High temperature

their resistance change with temperature into a voltage change. Figure 1-4 presents the temperature–resistance characteristics of common RTD sensors.

Optical pyrometers are utilized for temperature measurement when sensor physical contact with a process is not feasible but a view is available. Measurements are limited to energy emissions within the spectral response capability of the specific sensor used. A radiometric match of emissions between a calibrated reference source and the source of interest provides a current analog corresponding to temperature. Automatic pyrometers employ a servo loop to achieve this balance, as shown in Figure 1-5. Operation to 5000°C is available.

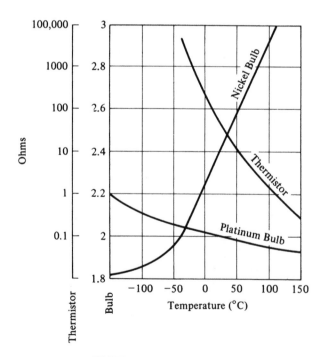

FIGURE 1-4. RTD devices.

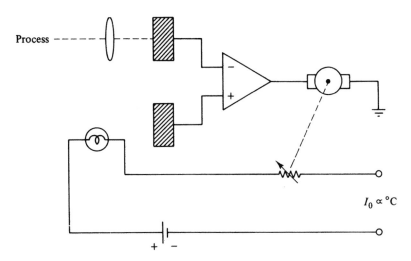

FIGURE 1-5. Automatic pyrometer.

1-3   MECHANICAL SENSORS

Fluid pressure is defined as the force per unit exerted by a gas or a liquid on the boundaries of a containment vessel. Pressure is a measure of the energy content of hydraulic and pneumatic (liquid and gas) fluids. Hydrostatic pressure refers to the internal pressure at any point within a liquid directly proportional to the liquid height above that point, independent of vessel shape. The static pressure of a gas refers to its potential for doing work, which does not vary uniformly with height as a consequence of its compressibility. Equation (1-3) expresses the basic relationship between pressure, volume, and temperature as the general gas law. Pressure typically is expressed in terms of pounds per square inch (psi) or inches of water (in H_2O) or mercury (in Hg). Absolute pressure measurements are referenced to a vacuum, whereas gauge pressure measurements are referenced to the atmosphere.

$$\frac{\text{Absolute temperature} \times \text{Gas volume}}{\text{Absolute pressure}} = \text{Constant} \qquad (1-3)$$

A pressure sensor detects pressure and provides a proportional analog signal by means of a pressure–force summing device. This usually is implemented with a mechanical diaphragm and linkage to an electrical element such as a potentiometer, strain gauge, or piezoresistor. Quantities of interest associated with pressure–force summing sensors include their mass, spring constant, and natural frequency. Potentiometric elements are low in cost and have high output, but their sensitivity to vibration and mechanical nonlinearities combine to limit their utility. Unbonded strain gauges offer improvement in accuracy and stability, with errors to 0.5% of full scale, but their low output signal requires a preamplifier. Present developments in

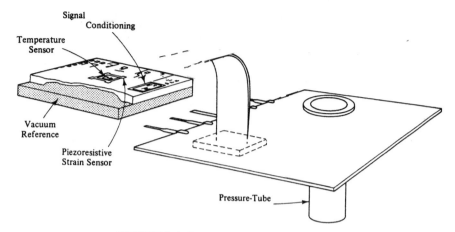

FIGURE 1-6. Integrated pressure microsensor.

pressure transducers involve integral techniques to compensate for the various error sources, including crystal diaphragms for freedom from measurement hysteresis. Figure 1-6 illustrates a microsensor circuit pressure transducer for enhanced reliability with an internal vacuum reference, chip heating to minimize temperature errors, and a piezoresistor bridge transducer circuit with on-chip signal conditioning.

Liquid levels are frequently required to process measurements in tanks, pipes, and other vessels. Sensing methods of various complexity are employed, including float devices, differential pressure, ultrasonics, and bubblers. Float devices offer simplicity and various ways of translating motion into a level reading. A differential pressure transducer can also measure the height of a liquid when its specific weight W is known, and a ΔP cell is connected between the vessel surface and bottom. Height is provided by the ratio of $\Delta P/W$.

Accurate sensing of position, shaft angle, and linear displacement is possible with the linear variable displacement transformer (LVDT). With this device, an ac excitation introduced through a variable reluctance circuit is induced in an output circuit through a movable core that determines the amount of displacement. LVDT advantages include overload capability and temperature insensitivity. Sensitivity increases with excitation frequency, but a minimum ratio of 10:1 between excitation and signal frequencies is considered a practical limit. LVDT variants include the induction potentiometer, synchros, resolvers, and the microsyn. Figure 1-7 describes a basic LVDT circuit with both ac and dc outputs.

Fluid flow measurement generally is implemented either by differential pressure or mechanical contact sensing. Flow rate F is the time rate of fluid motion with dimensions typically in feet per second. Volumetric flow Q is the fluid volume per unit time, such as gallons per minute. Mass flow rate M for a gas is defined, for example, in terms of pounds per second. Differential pressure flow sensing elements are also known as variable head meters because the pressure difference between the two measurements ΔP is equal to the head. This is equiv-

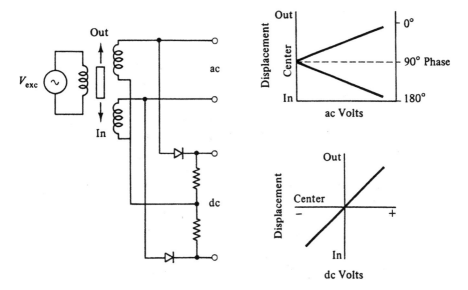

FIGURE 1-7. Basic LVDT.

alent to the height of the column of a differential manometer. Flow rate is therefore obtained with the 32 ft/sec^2 gravitational constant g and differential pressure by equation (1-4). Liquid flow in open channels is obtained by head-producing devices such as flumes and weirs. Volumetric flow is obtained with the flow cross-sectional area and the height of the flow over a weir, as shown by Figure 1-8 and equation (1-5).

$$\text{Flow rate } F = \sqrt{2g\Delta P} \text{ feet/second} \tag{1-4}$$

$$\text{Volumetric flow } Q = \sqrt{2gL^2H^3} \text{ cubic feet/second} \tag{1-5}$$

$$\text{Mass flow } M = \sqrt{R\frac{\Delta P_0}{\Delta P_x}} \cdot \sqrt{\frac{P\Delta P}{T}} \text{ pounds/second} \tag{1-6}$$

where
  R = universal gas constant
  ΔP_0 = true differential pressure, $P_0 - P_\infty$
  ΔP_x = calibration differential pressure

Acceleration measurements are principally of interest for shock and vibration sensing. Potentiometric dashpots and capacitive transducers have largely been supplanted by piezoelectric crystals. Their equivalent circuit is a voltage source in series with a capacitance, as shown in Figure 1-9 which produces an output in

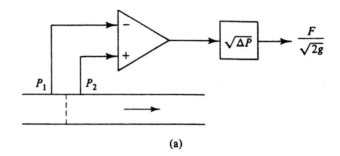

(a)

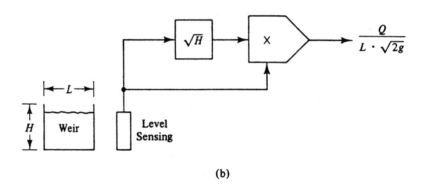

(b)

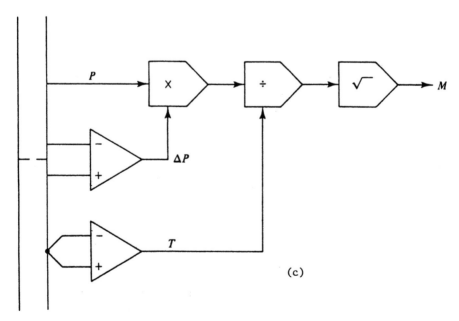

(c)

FIGURE 1-8. (a) Flow rate, (b) volumetric flow, and (c) mass flow.

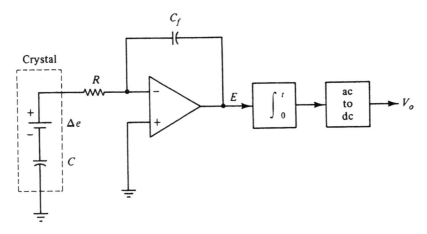

FIGURE 1-9. Vibration measurement.

coulombs of charge as a function of acceleration excitation. Vibratory acceleration results in an alternating output typically of very small value. Several crystals are therefore stacked to increase the transducer output. As a consequence of the small quantities of charge transferred, this transducer usually is interfaced to a low-input-bias-current charge amplifier, which also converts the acceleration input to a velocity signal. An ac-coupled integrator will then provide a displacement signal that may be calibrated, for example, in milliinches of displacement per volt, as shown in Figure 2-14.

A load cell is a transducer whose output is proportional to an applied force. Strain gauge transducers provide a change in resistance due to mechanical strain produced by a force member. Strain gauges may be based on a thin metal wire, foil, thin films, or semiconductor elements. Adhesive-bonded gauges are the most widely used, with a typical resistive strain element of 350 Ω that will register full-scale changes to 15 Ω. With a Wheatstone bridge circuit, a 2V excitation may therefore provide up to a 50 mV output signal change, as described in Figure 1-10. Semiconductor strain gauges offer high sensitivity at low strain levels, with outputs of 200 mV to 400 mV. Miniature tactile force sensors can also be fabricated from scaled-down versions of classic transducers employing MEMS technology. A multiplexed array of these sensors can provide sense feedback for robotic part manipulation and teleoperator actuators.

Ultrasound ranging and imaging systems are increasingly being applied for industrial and medical purposes. A basic ultrasonic system is illustrated by Figure 1-11; it consists of a phased array transducer and associated signal processing, including aperture focusing by means of time delays, and is employed in both medical ultrasound and industrial nondestructive testing applications. Multiple frequency emissions in the 1–10 MHz range are typically employed to prevent spatial multipath cancellations. B-scan ultrasonic imaging displays acoustic reflectivity for a fo-

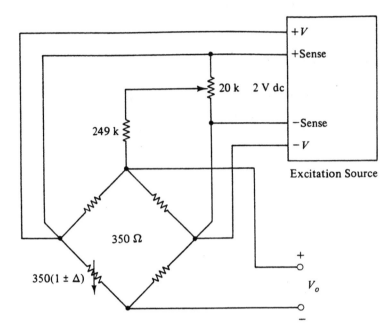

FIGURE 1-10. Strain gauge.

cal plane, and C-scan imaging provides integrated volumetric reflectivity of a region around the focal plane.

Hall effect transducers, which usually are silicon substrate devices, frequently include an integrated amplifier to provide a high-level output. These devices typically offer an operating range from –40 to +150°C and a linear output. Applications include magnetic field intensity sensing and position sensing with circuit isolation, such as the Micro Switch LOHET device, which offers a 3.75 mV/Gauss response. Figure 1-12 shows the principle of Hall effect operation. When a magnetic field B_z is applied perpendicular to a current-conducting element, a force acts on the current I_x, creating a diversion of its flow proportional to a difference of potential. This measurable voltage V_y is pronounced in materials such as InSb and InAs, and occurs to a useful degree in Si. The magnetic field usually is provided as a function of a measurand.

1-4 QUANTUM SENSORS

Quantum sensors are of significant interest as electromagnetic spectrum transducers over a frequency range extending from the far infrared region of 10^{11} Hz, through the visible spectrum about 10^{14} Hz, to the far ultraviolet region at 10^{17} Hz. These photon sensors are capable of measurements of a single photon whose energy E

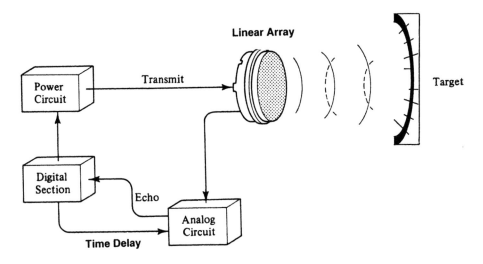

FIGURE 1-11. Phased array ultrasound system.

equals hv, or watt seconds in radiometry units from Table 1-3, where h is Planck's constant of 6.62×10^{-34} Joule seconds and v is frequency in Hz. Frequencies lower than infrared represent the microwave region and those higher than ultraviolet constitute X-rays, which require different transducers for measurement. In photometry, one lumen represents the power flux emitted over one steradian from a source of one candela intensity. For all of these sensors, incident photons result in an electrical signal by an intermediate transduction process.

Table 1-4 describes the relative performance of principal sensors. In photo diodes, photons generate electron–hole pairs within the junction depletion region. Photo transistors offer signal gain at the source for this transduction process ex-

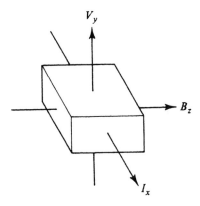

FIGURE 1-12. Hall effect transducer.

TABLE 1-3. Quantum Sensor Units

Parameter	Radiometry	Photometry	Photonic
Energy	Watt · sec	Lumen · sec	Photon
Irradiance	Watts/cm^2	Footcandles	Photon/sec/cm^2
Flux	Watts	Lumens	Photons/sec
Area radiance	$\dfrac{\text{Watts/steradian}}{\text{cm}^2}$	Footlamberts	Photon/sec/cm^2
Point intensity	Watts/steradian	Candelas · steradian	Photon/sec/steradian

TABLE 1-4. Sensor Relative Performance

Device	λ Region	$I_{photocurrent}/F_{photons/sec}$	Application
Photo diode	UV–near IR	10^{-3} amp/watt	Photonic detector
Photoconductive	Visible–near IR	1 amp/watt	Photo controller
Bolometer	Near IR–far IR	10^3 amps/watt	Superconducting IR
Photomultiplier	UV–near IR	10^6 amps/watt	Spectroscopy

ceeding that of the basic photo diode. In photoconductive cells, photons generate carriers that lower the sensor bulk resistance, but their utility is limited by a restricted frequency response. These sensors are shown in Figures 1-13 and 1-14. In all applications, it is essential to match sources and sensors spectrally in order to maximize energy transfer. For diminished photon sources, the photomultiplier excels, owing to a photoemissive cathode followed by high multiplicative gain to 10^6 from its cascaded dynode structure. The high gain and inherent low noise provided by coordinated multiplication results in a widely applicable sensor, except for the infrared region. Presently, the photomultiplier vacuum electron ballistic structure does not have a solid-state equivalent.

The measurement of infrared radiation is difficult as a result of the low energy of

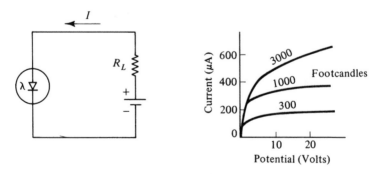

FIGURE 1-13. Photodiode characteristics.

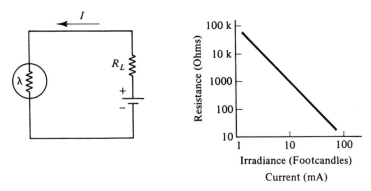

FIGURE 1-14. Photoconductive characteristics.

the infrared photon. This sensitivity deficiency has been overcome by thermally responsive resistive bolometer microsensors employing high-T_c superconductive films, whereby operation is maintained near the film transition temperature such that small temperature variations from infrared photons provide large resistance changes with gains to 10^3. Microsensor fabrication that enhances reliability and extension to arraying of elements is described in Figure 1-15, with image intensity $\bar{I} =$

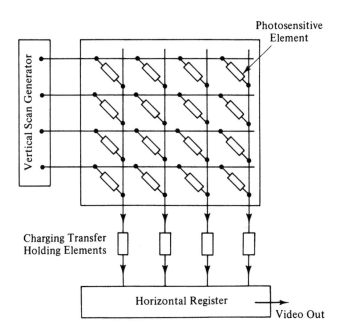

FIGURE 1-15. Quantum sensor array.

(x, y) quantized into a grey-level representation $f(\bar{I})$. This versatile imaging device is employed in applications ranging from analytical spectroscopy to night vision and space defense.

A property common to all nuclear radiation is its ability to interact with the atoms that constitute all matter. The nature of the interaction with any form of matter varies with the different components of radiation, as illustrated in Figure 1-16. These components are responsible for interactions with matter that generally produce ionization of the medium through which they pass. This ionization is the principal effect used in the detection of the presence of nuclear radiation. Alpha and beta rays often are not encountered because of their attenuation. Instruments for nuclear radiation detection are therefore most commonly constructed to measure gamma radiation and its scintillation or luminescent effect. The rate of ionization in Roentgens per hour is a preferred measurement unit, and represents the product of the emanations in Curies and in the sum of their energies in MeV represented as gamma energies. A distinction also should be made between disintegrations in counts per minute and ionization rate. The count rate measurement is useful for half-life determination and nuclear detection, but does not provide exposure rate information for interpretation of degree of hazard. The estimated yearly radiation dose to persons in the United States is 0.25 Roentgen (R). A high-radiation area is defined as one in which radiation levels exceed 0.1 R per hour, and requires posting of a caution sign.

Methods for detecting nuclear radiation are based on means for measuring the

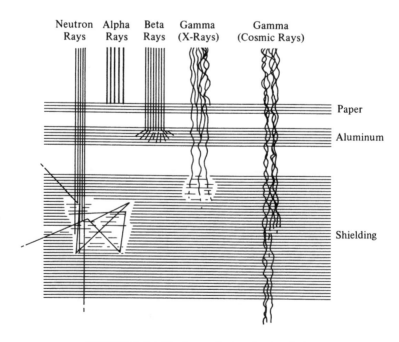

FIGURE 1-16. Nuclear radiation characteristics.

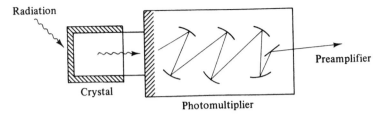

Radiation

Crystal

Photomultiplier

Preamplifier

FIGURE 1-17. Scintillation detector.

ionizing effects of these radiations. Mechanizations fall into the two categories of pulse-type detectors of ionizing events, and ionization-current detectors that employ an ionization chamber to provide an averaged radiation effect. The first category includes Geiger–Mueller tubes and more sensitive scintillation counters capable of individual counts. Detecting the individual ionizing scintillations is aided by an activated crystal such as sodium iodide optically coupled to a high-amplification photomultiplier tube, as shown in Figure 1-17. Ionization current detectors are primarily employed in health–physics applications such as industrial areas subject to high radiation levels. An ion chamber is followed by an amplifier whose output is calibrated in Roentgens per hour ionization rate. This method is necessary where pulse-type detectors are inappropriate because of a very high rate of ionization events. Practical industrial applications of nuclear radiation and detection include thickness gauges, nondestructive testing such as X-ray inspection, and chemical analysis such as by neutron activation.

1-5 ANALYTICAL SENSORS

A multisensor hierarchy representing signals common to the diverse instrumentation systems described throughout this text is shown in Figure 1-18. This perspective includes environmental sensors and actuators, such as temperature, pressure, level, and flow, applied at physical apparatus boundaries whose information content may be modeled by single-time constant transfer functions. In situ sensors provide more comprehensive measurements typically occurring beyond physical apparatus boundaries, such as capturing the dynamics of chemical reactions or evolving microstructure properties. These measurements frequently are multidimensional and modeled by multiinput–multioutput (MIMO) state variables.

The relationship between in situ and analytical measurements often is only a matter of sensor location; in situ measurements are acquired during real-time physical events, whereas analytical measurements may be acquired post-event, off-line, as an ex situ assay. Both real-time and post-event analytical measurements are encountered, however; for example, optical spectroscopy for the former and X-ray photon analysis (XPS) for the latter. The higher attribution of data at this level generally is expressed in terms of an ex situ feature model, also illustrated in Figure 1-18. These

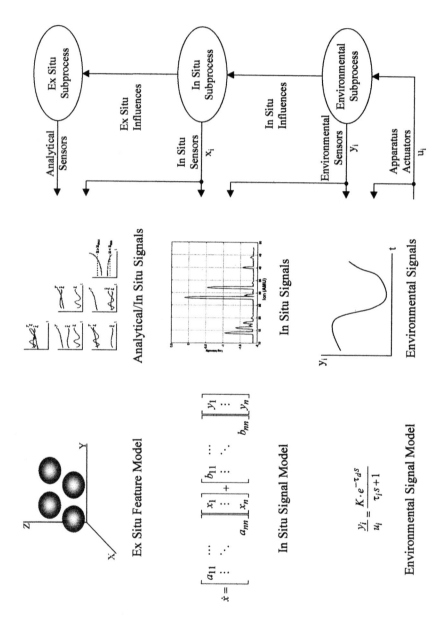

FIGURE 1-18. Multisensor signal model hierarchy.

18

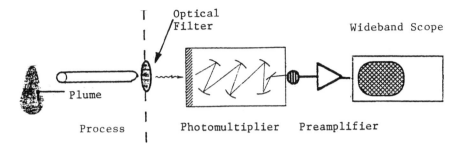

FIGURE 1-19. Optical spectrometer structure.

combined signal models provide useful describing functions that increasingly are applied as substitutes for conventional mathematical process models.

Chemical sensors are employed to determine the presence, concentration, or quantity of elemental or molecular analytes. These sensors may be divided into two classes: either electromagnetic, involving filtered optical and atomic mass unit spectroscopy; or electrochemical, involving the selectivity of charged species. Quantum spectroscopy, described by Figure 1-19, offers specific chemical measurements utilizing wavelength-selective filters from UV to near-IR coupled to a photoemissive photomultiplier whose output is displayed by a wide-band oscilloscope. Alternately, mass spectrometry chemical analysis is performed at high vacuum, typically employing a quadrupole filter shown in Figure 1-20, where sample gas is energized by an ion source. The mass filter selects ions determining specific charge-to-mass ratios, employing both electric and magnetic fields with the relationship $mV^2 = 2\ eV$, that are subsequently collected by an ion detector whose current intensity is displayed versus atomic mass unit (AMU).

Online measurements of industrial processes and chemical streams often require the use of selective chemical analyzers for the control of a processing unit. Examples include oxygen for boiler control, sulfur oxide emissions from combustion processes, and hydrocarbons associated with petroleum refining. Laboratory instruments such as gas chromatographs generally are not used for on-line measurements

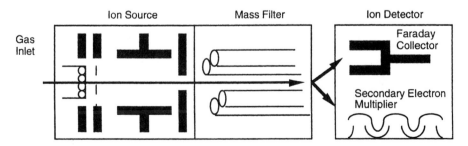

FIGURE 1-20. Mass spectrometer structure.

primarily because they analyze all compounds present simultaneously rather than a single one of interest.

The dispersive infrared analyzer is the most widely used chemical analyzer, owing to the range of compounds it can be configured to measure. Operation is by the differential absorption of infrared energy in a sample stream in comparison to that of a reference cell. Measurement is by deflection of a diaphragm separating the sample and reference cells, which in turn detunes an oscillator circuit to provide an electrical analog of compound concentration. Oxygen analyzers usually are of the amperometric type, in which oxygen is chemically reduced at a gold cathode, resulting in a current flow from a silver anode as a function of this reduction and oxygen concentration. In a paramagnetic wind device, a wind effect is generated when a mixture containing oxygen produces a gradient in a magnetic field. Measurement is derived by the thermal cooling effect on a heated resistance element forming a thermal anemometer. Table 1-5 describes basic electrochemical analyzer methods, and Figure 1-21 shows a basic gas analyzer system with calibration.

Also in this group are pH, conductivity, and ion-selective electrodes. pH defines the balance between the hydrogen ions H^+ of an acid and the hydroxyl ions OH^- of an alkali, where one type can be increased only at the expense of another. A pH probe is sensitive to the presence of H^+ ions in solution, thereby representing the acidity or alkalinity of a sample. All of these ion-selective electrodes are based on the Nernst equation (1-7), which typically provides a 60 mV potential change for each tenfold change in the activity of a monovalent ion.

$$V_0 = V + \frac{F}{n} \log(ac + s_1 a_1 c_1 + \ldots) \text{ volts} \qquad (1\text{-}7)$$

where
V_0 = voltage between sensing and reference electrodes
V = electrode base potential
F = Nernst factor, 60 mV at 25°C
n = ionic charge, 1 monovalent, 2 bivalent, etc.
a = ionic activity
c = concentration
s = electrode sensitivity to interfering ions

TABLE 1-5. Chemical Analyzer Methods

Compound	Analyzer
CO, SO_x, NH_x	Infrared
O_2	Amperometric, paramagnetic
HC	Flame ionization
NO_x	Chemiluminescent
H_2S	Electrochemical cell

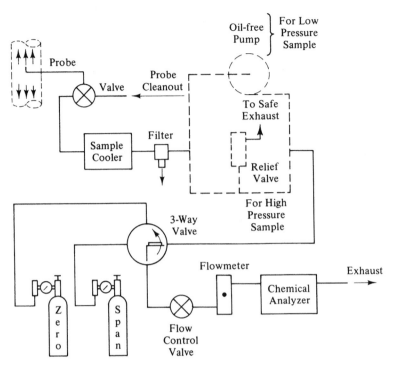

FIGURE 1-21. Calibrated gas analyzer.

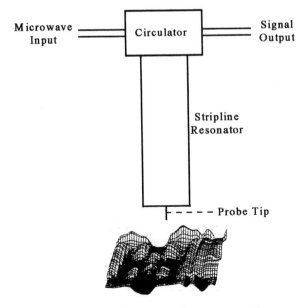

FIGURE 1-22. Microwave microscopy transducer.

22

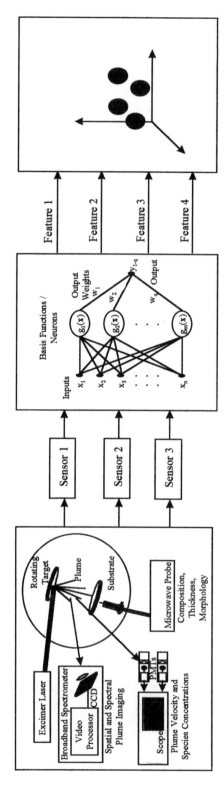

FIGURE 1-23. Multisensor multidimensional process features.

Evolving capabilities in digital imaging are extending process sensing by acquisition of data features employing spatial as well as spectral information. Techniques include microwave microscopy and electron wave functions such as scanning electron microscopy (SEM) that provide specimen topology details to submicrometer scale. In microwave microscopy, 30 GHz excitation is mixed with scanned specimen reflected energy, representing electrical permittivity and conductance changes, to acquire specimen topological details. This transducer is shown in Figure 1-22, and the complete instrument in Chapter 9, Section 9-3. Applications range from steel products subsurface defect detection, to high-Tc superconductor alloy evaluation, to biomedical assays. Multisensor architectures are described that enable improved data characterization and feature accessibility unavailable from single sensors, acquiring either homogeneous, alike information related to process parameters, or heterogeneous, different information to jointly account for process parameters, as described in Chapter 8. An example multisensor information structure is shown in Figure 1-23, employing computational feature extraction to achieve a higher attribution data presentation than is possible with single-sensor data.

BIBLIOGRAPHY

1. Petriu, E. M. (Ed.), *Instrumentation and Measurement Technology and Applications,* New York: IEEE, 1998.

2. Kovacs, G. T. A., *Micromachined Transducers Sourcebook,* New York: McGraw-Hill, 1998.

3. Gardner, J. W., *Microsensors,* New York: Wiley, 1994.

4. H. N. Norton, *Handbook of Transducers for Electronic Measuring Systems,* Englewood Cliffs, NJ: Prentice Hall, 1969.

5. Rabinovich, S. G., *Measurement Errors and Uncertainties,* 2nd ed., New York: Springer-Verlag, 1999.

6. Garrett, P. H. (contributing author), *Handbook of Industrial Automation,* New York: Marcel Dekker, 2000.

7. Prensky, S. D. and Castellucis, R. L., *Electronic Instrumentation,* 3rd ed., Englewood Cliffs, NJ: Prentice Hall, 1982.

8. Tabib-Azar, M. et al., "Super-Resolution Characterization of Microwave Conductivity of Semiconductors," *IOP Measurement Science Techn., 3,* 1993, pp. 583–590.

9. Jha, A. R., *Infrared Technology,* New York: Wiley, 2000.

10. Garrett, P. H. et al., "Emerging Methods for the Intelligent Processing of Materials," *Journal of Materials Engineering and Performance, 2, 5,* October 1993, pp. 727–732.

11. Laube, S. J. P. et al., "Sensor Principles and Methods," *Journal of Materials, 48,* 9, September 1996, pp. 16–23.

12. Garrett, P. H. et al., *Advanced Instrumentation and Computer I/O Design,* New York: IEEE Press, 1994.

13. Garrett, P. H. et al., "Self-Directed Processing of Materials," *IFAC, Engineering Applications of Artificial Intelligence,* Elsevier, 12, August 1999.

14. Breyfogle, F. W., Implementing Six Sigma, New York: Wiley-Interscience, 1999.

2

INSTRUMENTATION AMPLIFIERS AND PARAMETER ERRORS

2-0 INTRODUCTION

This chapter is concerned with the devices and circuits that comprise the electronic amplifiers of linear systems utilized in instrumentation applications. This development begins with the temperature limitations of semiconductor devices, and is then extended to differential amplifiers and an analysis of their parameters for understanding operational amplifiers from the perspective of their internal stages. This includes gain–bandwidth–phase stability relationships and interactions in multiple amplifier systems. An understanding of the capabilities and limitations of operational amplifiers is essential to understanding instrumentation amplifiers.

An instrumentation amplifier usually is the first electronic device encountered in a signal acquisition system, and in large part it is responsible for the ultimate data accuracy attainable. Present instrumentation amplifiers are shown to possess sufficient linearity, CMRR, low noise, and precision for total errors in the microvolt range. Five categories of instrumentation amplifier applications are described, with representative contemporary devices and parameters provided for each. These parameters are then utilized to compare amplifier circuits for implementations ranging from low input voltage error to wide bandwidth applications.

2-1 DEVICE TEMPERATURE CHARACTERISTICS

The elemental semiconductor device in electronic circuits is the *pn* junction; among its forms are diodes and bipolar and FET transistors. The availability of free carriers that result in current flow in a semiconductor is a direct function of the applied thermal energy. At room temperature, taken as 20°C (293°K above absolute zero), there is abundant energy to liberate the valence electrons of a semiconductor. These carriers are then free to drift under the influence of an applied potential. The magnitude of this current flow is essentially a function of the thermal energy instead of the applied voltage and accounts for the temperature behavior exhibited by semiconductor devices (increasing current with increasing temperature).

The primary variation associated with reverse biased *pn* junctions is the change in reverse saturation current I_s with temperature. I_s is determined by device geometry and doping with a variation of 7% per degree centigrade both in silicon and germanium, doubling every 10°C rise. This behavior is shown by Figure 2-1 and equation (2-1). Forward-biased *pn* junctions exhibit a decreasing junction potential, having an expected value of –2.0 mV per degree centigrade rise as defined by equation (2-2). The *dV/dT* temperature variation is shown to be the difference between the forward junction potential *V* and the temperature dependence of I_s. This relationship is the source of the voltage offset drift with temperature exhibited by semiconductor devices. The volt equivalent of temperature is an empirical model in both equations defined as $V_T = (273°K + T°C)/11{,}600$, having a typical value of 25 mV at room temperature.

$$\frac{dI_s}{dT} = I_s \cdot \frac{d(lnI_s)}{dT} \ A/°C \tag{2-1}$$

$$\frac{dV}{dT} = \left(\frac{V}{T} - \frac{V_T}{I_s} \cdot \frac{dI_s}{dT} \right) V/°C \tag{2-2}$$

2-2   DIFFERENTIAL AMPLIFIERS

The first electronic circuit encountered by a sensor signal in a data acquisition system typically is the differential input stage of an instrumentation amplifier. The balanced bipolar differential amplifier of Figure 2-2(a) is an important circuit used in many linear applications. Operation with symmetrical ± power supplies as shown results in the input base terminals being at 0 V under quiescent conditions. Due to the interaction that occurs in this emitter-coupled circuit, the algebraic difference

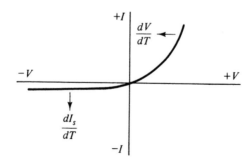

FIGURE 2-1. *pn* junction temperature dependence.

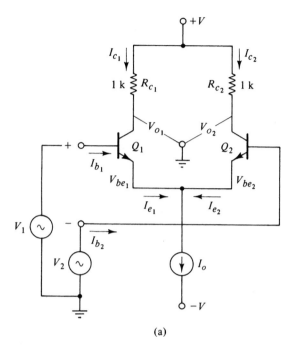

(a)

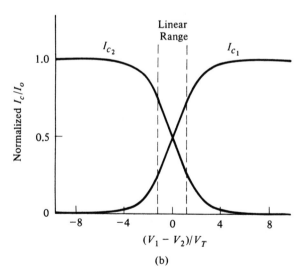

(b)

FIGURE 2-2. Differential DC amplifier and normalized transfer curves; $h_{fe} = 100$, $h_{ie} = 1$ k, and $h_{oe} = 10^{-6}$ ℧ .

signal applied across the input terminals is the effective drive signal, whereas equally applied input signals are cancelled by the symmetry of the circuit. With reference to a single-ended output V_{O_2}, amplifier Q_1 may be considered an emitter follower with the constant current source an emitter load impedance in the megohm range. This results in a noninverting voltage gain for Q_1 very close to unity (0.99999) that is emitter-coupled to the common emitter amplifier Q_2, where Q_2 provides the differential voltage gain $A_{V\text{diff}}$ by equation (2-3).

Differential amplifier volt–ampere transfer curves are defined by Figure 2-2(b), where the abscissa represents normalized differential input voltage $(V_1 - V_2)/V_T$. The transfer characteristics are shown to be linear about the operating point corresponding to an input voltage swing of approximately 50 mV (± 1 V_T unit). The maximum slope of the curves occurs at the operating point of $I_o/2$, and defines the effective transconductance of the circuit as $\Delta I_C/\Delta(V_1 - V_2)/V_T$. The value of this slope is determined by the total current I_o of equation (2-4). Differential input impedances R_{diff} and R_{cm} are defined by equations (2-5) and (2-6). The effective voltage gain cancellation between the noninverting and inverting inputs is represented by the common mode gain $A_{V\text{cm}}$ of equation (2-7). The ratio of differential gain to common mode gain also provides a dimensionless figure of merit for differential amplifiers as the common mode rejection ratio (CMRR). This is expressed by equation (2-8), having a typical value of 10^5.

$$A_{V\text{diff}} = \frac{h_{fe}R_c}{2h_{ie}} \quad \text{single-ended } V_{O_2} \tag{2-3}$$

$$= 50$$

$$I_o = I_{s_1} \cdot \exp(V_{be_1}/V_T) + I_{s_2} \cdot \exp(V_{be_2}/V_T) \tag{24}$$

$$= 1 \text{ mA}$$

$$R_{\text{diff}} = \frac{4V_T h_{fe}}{I_o} \tag{2-5}$$

$$= 10 \text{ K}$$

$$R_{\text{cm}} = \frac{hfe}{hoe} \tag{2-6}$$

$$= 100 \text{ M}$$

$$A_{V\text{cm}} = \frac{h_{oe}R_c}{2} \tag{2-7}$$

$$= 5 \times 10^{-4}$$

$$CMRR = \frac{A_{V\text{diff}}}{A_{V\text{cm}}} \tag{2-8}$$

$$= 10^5$$

The performance of operational and instrumentation amplifiers are largely determined by the errors associated with their input stages. It is convention to express these errors as voltage and current offset values, including their variation with temperature with respect to the input terminals, so that various amplifiers may be compared on the same basis. In this manner, factors such as the choice of gain and the amplification of the error values do not result in confusion concerning their true magnitude. It is also notable that the symmetry provided by the differential amplifier circuit primarily serves to offer excellent dc stability and the minimization of input errors in comparison with those of nondifferential circuits.

The base emitter voltages of a group of the same type of bipolar transistors at the same collector current are typically only within 20 mV. Operation of a differential pair with a constant current emitter sink as shown in Figure 2-2(a), however, provides a V_{be} match of V_{os} to about 1 mV. Equation (2-9) defines this input offset voltage and its dependence on the mismatch in reverse saturation current I_s between the differential pair. This mismatch is a consequence of variations in doping and geometry of the devices during their manufacture. Offset adjustment is frequently provided by the introduction of an external trimpot R_{Vos} in the emitter circuit. This permits the incremental addition and subtraction of emitter voltage drops to 0 V_{os} without disturbing the emitter current I_o.

Of greater concern is the offset voltage drift with temperature, dV_{os}/dT. This input error results from mistracking of V_{be_1} and V_{be_2}, described by equation (2-10), and is difficult to compensate. However, the differential circuit reduces dV_{os}/dT to 2 μV/°C from the –2 mV/°C for a single device of equation (2-2), or an improvement factor of 1/1000. By way of comparison, JFET differential circuit V_{os} is on the order of 10 mV, and dV_{os}/dT typically 5 μV/°C. Minimization of these errors is achieved by matching the device pinch-off voltage parameter. Bipolar input bias current offset and offset current drift are described by equations (2-11) and (2-12), and have their genesis in a mismatch in current gain ($h_{fe_1} \neq h_{fe_2}$). JFET devices intrinsically offer lower input bias currents and offset current errors in differential circuits, which is advantageous for the amplification of current-type sensor signals. However, the rate of increase of JFET bias current with temperature is exponential, as illustrated in Figure 2-3, and results in values that exceed bipolar input bias currents at temperatures beyond 100°C, thereby limiting the utility of JFET differential amplifiers above this temperature.

$$V_{os} = V_T \ln \frac{I_{s2}}{I_{s1}} \cdot \frac{I_{e1}}{I_{e2}} \tag{2-9}$$

$$= 1 \text{ mV}$$

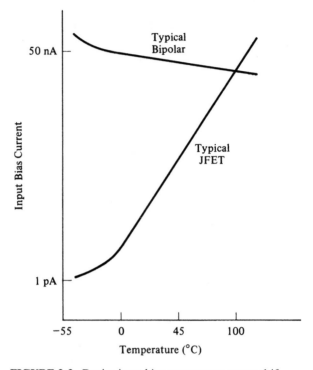

FIGURE 2-3. Device input bias current temperature drift.

$$\frac{dV_{os}}{dT} = \frac{dV_{be_1}}{dT} - \frac{dV_{be_2}}{dT} \tag{2-10}$$

$$= 2 \ \mu V/^\circ C$$

$$I_{os} = I_{b_1} - I_{b_2} \tag{2-11}$$

$$= 50 \ nA$$

$$\frac{dI_{os}}{dT} = B \cdot I_{os} \tag{2-12}$$

$$= 0.25 \ nA/^\circ C$$

$$B = -0.005/^\circ C > 25^\circ C$$

$$= -0.015/^\circ C < 25^\circ C$$

2-3   OPERATIONAL AMPLIFIERS

Most operational amplifiers are of similar design, as described by Figure 2-4, and consist of a differential input stage cascaded with a high-gain inner stage followed by a power output stage. Operational amplifiers are characterized by very high gain at dc and a uniform rolloff in this gain with frequency. This enables these devices to accept feedback from arbitrary networks with high stability and simultaneous dc and ac amplification. Consequently, such networks can accurately impart their characteristics to electronic systems with negligible degradation. The earliest integrated circuit amplifier was offered in 1963 by Texas Instruments, but the Fairchild 709 introduced in 1965 was the first operational amplifier to achieve widespread application. Improvements in design resulted in second-generation devices such as the National LM108. Advances in fabrication technology made possible amplifiers such as by the Analog Devices OP-07, with improved performance overall. Subsequent refinements are represented by devices including the Linear LTC-1250, featuring zero drift and ultralow noise. It is notable that contemporary operational amplifier circuits are structured around a high-gain inner-stage employing a constant current source active load. The gain stage active load impedance of approximately 500 K ohms ratioed with an emitter resistance R_e approximating 100 ohms, shown in Figure 2-4, is responsible for high overall A_{V_O}.

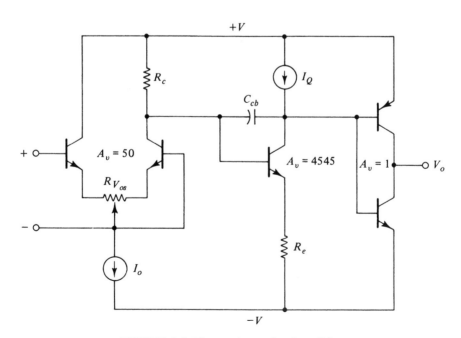

FIGURE 2-4. Elemental operational amplifier.

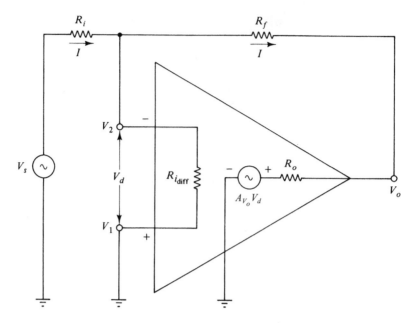

FIGURE 2-5. Inverting operational amplifier.

Since

$$R_{\text{diff}} \to \infty, \ V_d = \frac{V_o}{A_{V_o}} \to 0 \text{ as } |A_{V_o}| \to \infty$$

$$A_{V_c} = \frac{V_o}{V_s} = \frac{-IR_f}{IR_i} = \frac{-R_f}{R_i} \tag{2-13}$$

The circuit for an inverting operational amplifier is shown in Figure 2-5. The cascaded innerstage gains of Figure 2-4 provide a total open-loop gain A_{V_o} of 227,500, enabling realization of the ideal closed-loop gain A_{V_c} representation of equation (2-13). In practice, the A_{V_o} value cannot be utilized without feedback because of nonlinearities and instability. The introduction of negative feedback between the output and inverting input also results in a virtual ground with equilibrium current conditions maintaining $V_d = V_1 - V_2$ at zero. Classification of operational amplifiers is primarily determined by the active devices that implement the amplifier differential input. Table 2-1 delineates this classification.

According to negative feedback theory, an inverting amplifier will be unstable if its gain is equal to or greater than unity when the phase shift reaches $-180°$ through the amplifier. This is so because an output-to-input relationship will also have been established, providing an additional $-180°$ by the feedback network. The relationships between amplifier gain, bandwidth, and phase are described by Figure 2-6 and

TABLE 2-1. Operational Amplifier Types

Bipolar	Prevalent type used for a wide range of signal processing applications. Good balance of performance characteristics.
FET	Very high input impedance. Frequently employed as an instrumentation-amplifier preamplifier. Exhibits larger input errors than bipolar devices.
CAZ	Bipolar device with auto-zero circuitry for internally measuring and correcting input error voltages. Provides low-input-uncertainty amplification.
BiFET	Combined bipolar and FET circuit for extended performance. Intended to displace bipolar devices in general-purpose applications.
Superbeta	A bipolar device approaching FET input impedance with the lower bipolar errors. A disadvantage is lack of device ruggedness.
Micropower	High-performance operation down to 1 volt supply powered from residual system potentials. Employs complicated low-power circuit equivalents for implementation.
Isolation	An internal barrier device using modulation or optical methods for very high isolation. Medical and industrial applications.
Chopper	dc–ac–dc circuit with a capacitor-coupled internal amplifier providing very low input voltage offset errors for minimum input uncertainty.
Varactor	Varactor diode input device with very low input bias currents for current amplification applications such as photomultipliers.
Vibrating capacitor	A special input circuit arrangement requiring ultralow input bias currents for applications such as electrometers.

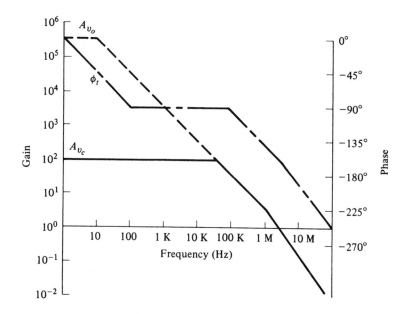

FIGURE 2-6. Operational amplifier gain–bandwidth–phase relationships.

equations (2-14) through (2-16) for an example closed-loop gain A_{V_c} value of 100. Each discrete inner stage contributes a total of $-90°$ to the cumulative phase shift ϕ_t , with $-45°$ realized at the respective -3 dB frequencies. The high-gain stage -3 dB frequency of 10 Hz is attributable to the dominant-pole compensating capacitance C_{cb} shown in Figure 2-4. The second corner frequency at 1 MHz is typical for a differential input stage, and the third at 25 MHz is contributed by the output stage. The overall phase margin of $30°$ $(180° - \phi_t)$ at the A_{V_c} unity gain crossover frequency of 2 MHz insures unconditional stability and freedom from a ringing output response.

$$A_{V_o} = \frac{227,250}{\left(1 + j\dfrac{f}{10\ \text{Hz}}\right)\left(1 + j\dfrac{f}{1\ \text{MHz}}\right)\left(1 + j\dfrac{f}{25\ \text{MHz}}\right)} \tag{2-14}$$

$$\phi_t = -\tan^{-1}\left(\frac{f}{10\ \text{Hz}}\right) - \tan^{-1}\left(\frac{f}{1\ \text{MHz}}\right) - \tan^{-1}\left(\frac{f}{25\ \text{MHz}}\right) \tag{2-15}$$

$$\text{Phase margin} = 180° - \phi_t \tag{2-16}$$

2-4   INSTRUMENTATION AMPLIFIERS

The acquisition of accurate measurement signals, especially low-level signals in the presence of interference, requires amplifier performance beyond the typical signal acquisition capabilities of operational amplifiers. An instrumentation amplifier is usually the first electronic device encountered by a sensor in a signal acquisition channel, and in large part it is responsible for the ultimate data accuracy attainable. Present instrumentation amplifiers possess sufficient linearity, stability, and low noise for total error in the microvolt range, even when subjected to temperature variations, on the order of the nominal thermocouple effects exhibited by input lead connections. High CMRR is essential for achieving the amplifier performance of interest with regard to interference rejection, and for establishing a signal ground reference at the amplifier that can accommodate the presence of ground return potential differences. High amplifier input impedance is also necessary to preclude input signal loading and voltage divider effects from finite source impedances, and to accommodate source impedance imbalances without degrading CMRR. The precision gain values possible with instrumentation amplifiers, such as 1000.000, are equally important to obtain accurate scaling and registration of measurement signals.

The relationship of CMRR to the output signal V_o for an operational or instrumentation amplifier is described by equation (2-17), and is based on the derivation of CMRR provided by equation (2-8). For the operational amplifier subtractor circuit of Figure 2-7, $A_{V\text{diff}}$ is determined by the feedback-to-input resistor ratios (R_f/R_i, with practically realizable values to 10^2, and $A_{V\text{cm}}$ is determined by the mis-

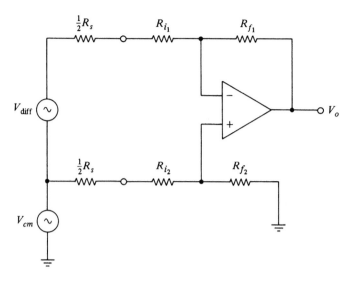

FIGURE 2-7. Subtractor instrumentation amplifier.

match between feedback and input resistor values attributable to their tolerances. Consequently, the $A_{V\text{cm}}$ for a subtractor circuit may be obtained from equation (2-18) and as tabulated in Table 2-2 to determine the average expected CMRR value for specified resistor tolerances. Notice that CMRR increases with $A_{V\text{diff}}$ by the numerator of equation (2-8), but $A_{V\text{cm}}$ is constant because of its normalization by the resistor tolerance chosen.

$$V_o = A_{V\text{diff}} \cdot V_{\text{diff}} + A_{V\text{cm}} \cdot V_{\text{cm}} \tag{2-17}$$

$$= A_{V\text{diff}} \cdot V_{\text{diff}}\left(1 + \frac{1}{\text{CMRR}} \cdot \frac{V_{\text{cm}}}{V_{\text{diff}}}\right)$$

$$\text{CMRR}_{\text{subtractor}} = \frac{\dfrac{1}{2}\left(\left|\dfrac{R_{f2} \pm \Delta R_{f2}}{R_{i2} \pm \Delta R_{i2}}\right| + \left|\dfrac{R_{f1} \pm \Delta R_{f1}}{R_{i1} \pm \Delta R_{i1}}\right|\right)}{\left(\left|\dfrac{R_{f2} \pm \Delta R_{f2}}{R_{i2} \pm \Delta R_{Ri2}}\right| - \left|\dfrac{R_{f1} \pm \Delta R_{f1}}{R_{i1} \pm \Delta R_{i1}}\right|\right)} \tag{2-18}$$

TABLE 2-2. Subtractor CMRR Expected Values

Resistor Tolerance	5%	2%	1%	0.1%
$A_{V\text{cm subtractor}}$	0.1	0.04	0.02	0.002
$\text{CMRR}_{\text{subtractor}}$ $(x A_{V\text{diff}})$	10	25	50	500

The subtractor circuit is capable of typical values of CMRR to 10^4, and its implementation is economical owing to the requirement for a single operational amplifier. However, its specifications are usually marginal when compared with the requirements of typical signal acquisition applications. For example, each implementation requires the matching of four resistors, and the input impedance is constrained to the value of R_i chosen. For modern bipolar amplifiers, such as the Analog Devices OP-07 and Burr Brown OPA-128 devices with gigohm internal resistances, megohm R_i values are allowable to prevent input voltage divider effects resulting from an imbalanced kilohm R_s source resistance. Further, low-bias-current amplifiers are essential for current sensors including nuclear gauges, pH probes, and photomultiplier tubes. The OPA-128 also offers a balance of input parameters for this application with an I_{os} of 30 fA and typical current sensor R_s values of 10 M ohms. The compensating resistor R_c shown in Figure 2-8 is matched to R_s in order to preserve CMRR. The five amplifiers presented in Table 2-3 beneficially permit the comparison of limiting parameters that influence performance in specific amplifier applications, where the CMRR entries described are expected in-circuit values.

The three-amplifier instrumentation amplifier of Figure 2-9, exemplified by the AD624, offers improved performance overall compared to the foregoing subtractor circuit with in-circuit $CMRR_{3ampl}$ values of 10^5 and the absence of problematic external discrete input resistors. In order to minimize output noise and offsets with this amplifier, its subtractor A_{Vdiff} is normally set to unity gain. The first stage of this amplifier also has a unity A_{Vcm}, owing to its differential-input-to-differential-output connection, which results in identical first-stage CMRR and A_{Vdiff} values.

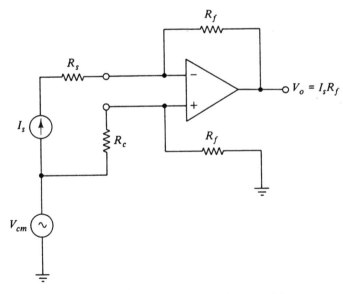

FIGURE 2-8. Differential current-voltage amplifier.

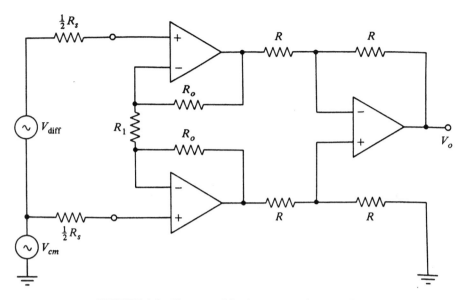

FIGURE 2-9. Three-amplifier instrumentation amplifier.

Amplifier internal resistance trimming consequently achieves the nominal subtractor A_{Vcm} value shown in equation (2-19).

The differential output instrumentation amplifier, illustrated by Figure 2-10, offers increased common mode rejection via equation (2-20) over the three-amplifier circuit from the addition of a second output subtractor. By comparison, a single subtractor permits a full-scale 24 V_{pp} output signal swing, whereas dual subtractors deliver a full-scale 48 V_{pp} output signal from opposite polarity swings of the ±15 V dc power supplies for each signal half cycle. The effective output gain doubling combined with first-stage gain provides $CMRR_{diff}$ output values to 10^6. This advanced amplifier circuit permits high-performance analog signal acquisition and the continuation of common mode interference rejection over a signal transmission channel, with termination by a remote differential-to-single-ended subtractor amplifier.

$$CMRR_{3\,ampl} = CMRR_{1st\,stage} \cdot CMRR_{subtractor} \qquad (2-19)$$

$$= \frac{A_{Vdiff\,1st\,stage}}{1} \cdot \frac{1}{A_{Vcm\,subtractor}}$$

$$= \left(1 + \frac{2R_0}{R_1}\right) \cdot \left(\frac{1}{0.001}\right)$$

$$CMRR_{diff\,output} = \left(1 + \frac{2R_0}{R_G}\right) \cdot \left(\frac{2}{0.001}\right) \qquad (2-20)$$

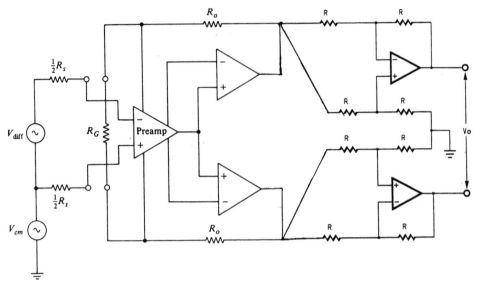

FIGURE 2-10. Differential output instrumentation amplifier.

Isolation amplifiers are advantageous for very noisy and high-voltage environments plus the interruption of ground loops. They further provide galvanic isolation typically on the order of 1 μA input-to-output leakage. The front end of the isolation amplifier is similar to an instrumentation amplifier, as shown in Figure 2-11, and is operated from an internal dc–dc isolated power supply to insure isolation integrity and for external sensor excitation purposes. As a consequence, these amplifiers do not require sourcing or sinking external bias currrent, and function normally with fully floating sensors. Most designs also include a 100 K ohm series input resistor R to limit catastrophic fault currents. Typical isolation barriers have an equivalent circuit of 10^{11} ohms shunted by 10 pF, representing R_{iso} and C_{iso}. An input-to-output V_{iso} rating of 1500 V rms is common, and has a corollary isolation mode rejection ratio (IMRR) with reference to the output. CMRR values of 10^5 relative to the input are common, and IMRR values to 10^8 with reference to the output are available at 60 Hz. This capability makes possible the accommodation of two sources of interference, V_{cm} and V_{iso}, both frequently encountered in sensor applications. The performance of this connection is described by equation (2-21).

$$V_o = A_{Vdiff} \cdot V_{diff}\left(1 + \frac{1}{\text{CMRR}} \cdot \frac{V_{cm}}{V_{diff}}\right) + \frac{V_{iso}}{\text{IMRR}} \qquad (2\text{-}21)$$

High-speed data conversion and signal conditioning circuits capable of accommodating pulse and video signals require wideband operational amplifiers. Such amplifiers are characterized by their settling time, delay, slew rate, and transient subsidence, described in Figure 2-12. Parasitic reactive circuit elements and care-

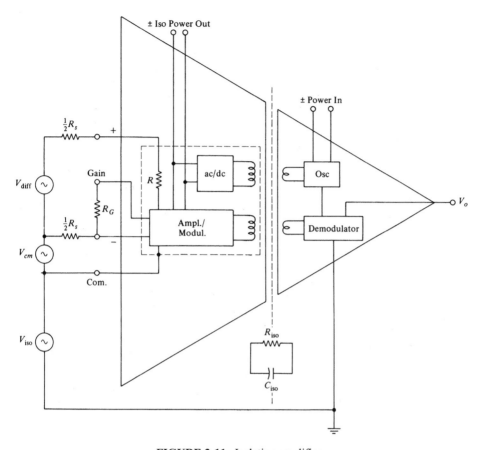

FIGURE 2-11. Isolation amplifier.

lessly planned circuit layouts result in performance derogation. Amplifier slew rate depends directly upon the product of the output voltage amplitude and signal frequency, and this product cannot exceed the slew rate specification of an amplifier if linear performance is to be realized. For example, a 1 V_{pp} sine wave signal at a frequency of 3 MHz typically encountered in video systems specifies an amplifier slew rate of at least 9.45 V/μs. If the amplifier is also loaded by 1000 pF of capacitance, then it must also be capable of delivering 10 mA of current output at that frequency. These relationships are described by equation (2-22) and its nomograph of Figure 2-13.

$$S_r = V_{opp} \cdot \pi \cdot f_{signal} \tag{2-22}$$

$$= \frac{I_o}{c_{sh}} \text{ V/s}$$

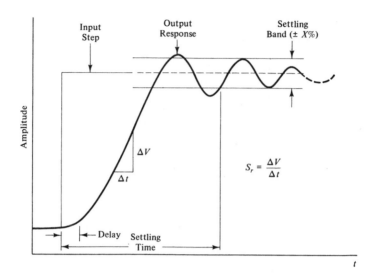

FIGURE 2-12. Wideband amplifler settling characteristics.

Acceleration sensors are principally of interest for shock and vibration measurements. Piezoelectric devices are prevalent transducers in this application and employ an equivalent circuit of a voltage source in series with a capacitive element, as shown in Figure 2-14, providing charge transfer as a function of acceleration mechanical inputs. A consequence of the small charge quantities transferred is the requirement for a low-bias-current amplifier whose circuit also converts acceleration

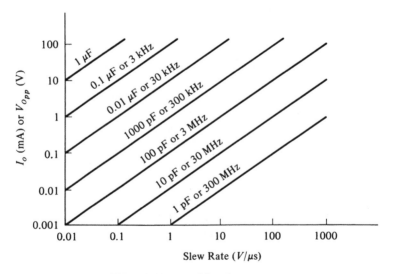

FIGURE 2-13. Amplifier slew rate curves.

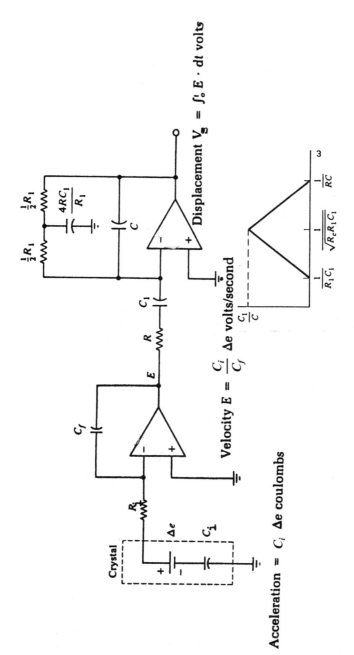

FIGURE 2-14. Accelerometer displacement ac integrator.

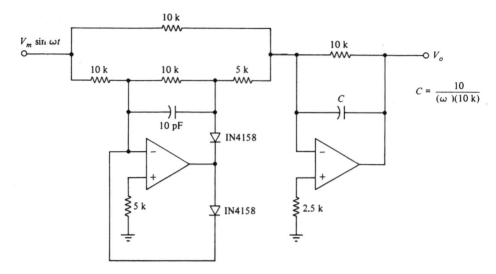

FIGURE 2-15. Precision ac-to-dc converter.

inputs into velocity signals. A following ac integrator provides a displacement output that may be calibrated, for example, in milliinches per volt. Accurate integration of very low frequency signals without saturation is possible owing to attenuation provided by the $1/R_1C_1$ cutoff frequency choice. Extended frequency differentiation is also available without noise sensitivity in this circuit by choice of the $1/RC$ upper cutoff. The circuit of Figure 2-15 enables accurate rectification of signals to submillivolt levels by employing active ac-to-dc conversion. This offers a conversion error of 0.6%FS with an RC filter cut-off of one-tenth the input signal frequency ω, and reduced error for larger RC products at the expense of additional response lag.

2-5 AMPLIFIER PARAMETER ERROR EVALUATION

The selection of an instrumentation amplifier involves the choice of amplifier input parameters that minimize amplification errors for applications of interest. It is therefore instructive to perform an error comparison between the five diverse amplifier types listed in Table 2-3, considering application-specific V_{cm} and R_s input values, with evaluation of voltage offsets, interference rejection, and gain nonlinearity. The individual error totals tabulated in Table 2-4 provide a performance summary expressed both as a referred-to-input (RTI) amplitude threshold uncertainty in volts, and as a percent of the full-scale output signal V_{oFS} following amplification by A_{Vdiff}. Error totals are derived from respective amplifier input parameter contributions defined in equation (2-23), where barred quantities denote mean values and unbarred quantities systematic and random values combined as the root-

TABLE 2-3. Amplifier Input Parameters Define Interface Applications

Symbol	Low Offset Voltage OP-07	Low Bias Current OPA-128	Three-Amplifier Instrumentation AD624	High-Voltage Isolation AD215	Wideband Video OPA-646	Comment
V_{OS}	10 μV	140 μV	25 μV	0.4 mV	1 mV	Offset voltage
$\frac{dV_{OS}}{dT}$	0.2 μV/°C	5 μV/°C	0.25 μV/°C	2 μV/°C	12 μV/°C	Offset voltage drift
I_{OS}	0.3 nA	30 fA	10 nA	300 nA	0.4 μA	Offset current
$\frac{dI_{OS}}{dT}$	5 pA/°C	Negligible	20 pA/°C	1 nA/°C	10 nA/°C	Offset current drift
Sr	0.3 V/μs	3 V/μs	5 V/μs	6 V/μs	180 V/μs	Slew rate
f_{hi}	600 KHz	500 KHz	1 MHz	120 KHz	650 MHz	Unity gain bandwidth
CMRR	10^4	10^4	10^5	10^5	10^4	Av_{diff}/Av_{cm}
V_{CM}	10 V rms	10 V rms	10 V rms	1500 V rms	10 V rms	Maximum applied volts
V_n rms	10 nV/$\sqrt{Hz}$	27 nV/$\sqrt{Hz}$	4 nV/$\sqrt{Hz}$	Negligable	7.5 nV/$\sqrt{Hz}$	Voltage noise
$f(A_v)$	0.01%	0.01%	0.001%	0.0005%	0.025%	Gain nonlinearity
$\frac{dA_v}{dT}$	50 ppm/°C	50 ppm/°C	5 ppm/°C	15 ppm/°C	50 ppm/°C	Gain drift
R_{diff}	8×10^7 Ω	10^{13} Ω	10^9 Ω	10^{12} Ω	15 KΩ	Differential resistance
R_{cm}	2×10^{11} Ω	10^{13} Ω	10^9 Ω	5×10^9 Ω	1.6 MΩ	Common mode resistance

TABLE 2-4. Amplifier Error Comparison ($V_{\text{diff}} = 1$ V, $A_{V\text{diff}} = 1$, $V_{o\text{FS}} = 1$ V, $dT = 10°C$)

	OP-07	OPA-128	AD624	AD215	OPA-646	Comment
R_s V_{CM}	10 KΩ ±10 V	10 MΩ ±10 V	1 KΩ ±10 V	50 Ω ±1000 V	75 Ω ±10 V	Input group
V_{OS}	$\overline{10}$ µV	$\overline{140}$ µV	$\overline{25}$ µV	$\overline{400}$ µV	$\overline{1000}$ µV	
$\dfrac{dV_{\text{OS}}}{dT} \cdot dT$	2 µV	50 µV	2.5 µV	20 µV	120 µV	Offset group
$I_{\text{OS}} \cdot R_s$	$\overline{3}$ µV	$\overline{0.3}$ µV	$\overline{10}$ µV	$\overline{15}$ µV	$\overline{30}$ µV	
$6.6\, V_n \sqrt{f_{\text{hi}}}$	51 µV	126 µV	26 µV	Negligible	1262 µV	
$\dfrac{V_{\text{CM}}}{\text{CMRR}}$	1000 µV	1000 µV	100 µV	10,000 µV	1000 µV	Interference group
$f(A_v) \cdot \dfrac{V_{o\text{FS}}}{A_{V\text{diff}}}$	$\overline{100}$ µV	$\overline{100}$ µV	$\overline{10}$ µV	$\overline{50}$ µV	$\overline{250}$ µV	Nonlinearity group
$\dfrac{dA_V}{dT} \cdot dT \cdot \dfrac{V_{o\text{FS}}}{A_{V\text{diff}}}$	500 µV	500 µV	50 µV	150 µV	500 µV	
$\varepsilon_{\text{ampl RTI}}$	$(\overline{113} + 1119)$ µV	$(\overline{240} + 1126)$ µV	$(\overline{45} + 115)$ µV	$(\overline{465} + 10{,}003)$ µV	$(\overline{1280} + 1690)$ µV	$\Sigma\ \overline{\text{mean}} + \text{RSS}_{\text{other}}$
$\varepsilon_{\text{ampl}\%\text{FS}}$	0.123%FS	0.136%FS	0.016%FS	1.046%FS	0.297%FS	$\dfrac{A_{V\text{diff}}}{V_{o\text{FS}}} \cdot 100\%$

sum-square. Note that $A_{V\text{diff}}$ normally is scaled for the V_{diff} input signal maximum in order to achieve a $V_{o\text{FS}}$ of interest at the amplifier output. However, for the normalized examples of Table 2-4, each $A_{V\text{diff}}$ is unity, requiring input V_{diff} values that equal the $V_{o\text{FS}}$ value.

$$\varepsilon_{\text{ampl\%FS}} = \{\varepsilon_{\text{amplRTI volts}}\} \times \frac{A_{V\text{diff}}}{V_{o\text{FS}}} \cdot 100\% \qquad (2\text{-}23)$$

$$= \left\{\overline{V_{\text{OS}} + \overline{I_{\text{OS}} \cdot R_S} + f\!\left(A_V \cdot \frac{V_{o\text{FS}}}{A_{V\text{diff}}}\right)}\right.$$

$$+ \left[\left(\frac{dV_{\text{OS}}}{dT} \cdot dT\right)^2 + \left(\frac{V_{\text{CM}}}{\text{CMRR}}\right)^2 + (6.6\, V_n\, \sqrt{f_{\text{hi}}})^2\right.$$

$$\left.\left.+ \left(\frac{dA_v}{dT} \cdot dT \cdot \frac{V_{o\text{FS}}}{A_{V\text{diff}}}\right)^2\right]^{1/2}\right\} \times \frac{A_{V\text{diff}}}{V_{o\text{FS}}} \cdot 100\%$$

   Each amplifier is evaluated at identical $A_{V\text{diff}}$, $V_{o\text{FS}}$, and temperature dT for consistency, but at expected R_s and V_{cm} values relevant to their typical application. All of the amplifiers are capable of accommodating off-ground and electromagnetically coupled V_{cm} input interference with an effectiveness determined by their respective incircuit CMRR, where the influence of amplifier CMRR values in attenuating respective V_{cm} values is described. Mean offset voltages V_{OS} are also untrimmed to reveal these possible differences. The OP-07 is assumed applied to an austere, four-resistor subtractor circuit resulting in its 10K Ω R_s, whereas the OPA-128 low-input bias current amplifier interfaces a 10M Ω R_s current sensor. The AD624 three-amplifier circuit offers the best performance and robustness overall, with gain nonlinearity values a tenth that of the other amplifiers, all of which are normalized to amplifier inputs by the ratio $V_{o\text{FS}}/A_{V\text{diff}}$.
   The AD215 isolation amplifier 50Ω R_s represents either the output of a preceding front-end instrumentation amplifier or low-level emf sensor. It is notable that the presence of a 1000 volt V_{cm} input essentially accounts for the total error of this amplifier, which will be safely accommodated by the amplifier physical structure. Finally, with a 75 Ω R_s, the wideband OPA-646 differs from other amplifiers in providing 10 times the bandwidth at 10 times the internal noise contribution. All of the amplifier error totals are commensurable owing to like manufacturing technologies. Amplifier V_n rms internal noise voltage is converted to peak–peak with multiplication by 6.6, to account for its crest factor, dimensionally equivocating it to other amplifier input values in each error total.

BIBLIOGRAPHY

1. G. Tobey, J. Graeme, and L. Huelsman, *Operational Amplifiers: Design and Applications,* New York: McGraw-Hill, 1971.

2. J. Embinder, *Application Considerations for Linear Integrated Circuits,* New York: Wiley-Interscience, 1970.

3. G. B. Rutkowski, *Handbook of Integrated-Circuit Operational Amplifiers,* Englewood Cliffs, NJ: Prentice Hall, 1975.

4. F. C. Fitchen, *Electronic Integrated Circuits and Systems,* New York: Van Nostrand Reinhold, 1970).

5. J. A. Connelly, *Analog Integrated Circuits,* New York: Wiley-Interscience, 1975.

6. J. M. Pettit and M. M. McWhorter, *Electronic Amplifier Circuits,* New York: McGraw-Hill, 1961.

7. J. G. Graeme, *Applications of Operational Amplifiers: Third-Generation Techniques,* New York: McGraw-Hill, 1973.

8. P. H. Garrett, *Analog I/O Design, Acquisition: Conversion: Recovery,* Reston, VA: Reston Publishing Co., 1981.

9. D. C. Bailey, "An Instrumentation Amplifier Is Not an Op Amp," *Electronic Products,* September 18, 1972.

10. J. W. Jaquay, "Designer's Guide to Instrumentation Amplifiers," *Electronic Design News,* May 5, 1972.

11. J. H. Kollataj, "Reject Common-Mode Noise," *Electronic Design,* April 26, 1973.

12. T. C. Lyerly, "Instrumentation Amplifier Conditions Computer Inputs," *Electronics,* November 6, 1972.

13. F. Poulist, "Simplify Amplifier Selection," *Electronic Design,* August 2, 1973.

14. Y. Netzer, "The Design of Low-Noise Amplifiers," *Proceedings IEEE,* June 1981.

15. H. W. Ott, *Noise Reduction Techniques in Electronic Systems,* New York: Wiley, 1976.

3

ACTIVE FILTER DESIGN WITH NOMINAL ERROR

3-0 INTRODUCTION

Although electric wave filters have been used for over a century since Marconi's radio experiments, the identification of stable and ideally terminated filter networks has occurred only during the past 35 years. Filtering at the lower instrumentation frequencies has always been a problem with passive filters because the required L and C values are larger and inductor losses appreciable. The band-limiting of measurement signals in instrumentation applications imposes the additional concern of filter error additive to these measurement signals when accurate signal conditioning is required. Consequently, this chapter provides a development of lowpass and bandpass filter characterizations appropriate for measurement signals, and develops filter error analyses for the more frequently required lowpass realizations.

The excellent stability of active filter networks in the dc to 100 kHz instrumentation frequency range makes these circuits especially useful. When combined with well-behaved Bessel or Butterworth filter approximations, nominal error band-limiting functions are realizable. Filter error analysis is accordingly developed to optimize the implementation of these filters for input signal conditioning, aliasing prevention, and output interpolation purposes associated with data conversion systems for dc, sinusoidal, and harmonic signal types. A final section develops maximally flat bandpass filters for application in instrumentation systems.

3-1 LOWPASS INSTRUMENTATION FILTERS

Lowpass filters are frequently required to band-limit measurement signals in instrumentation applications to achieve a frequency-selective function of interest. The application of an arbitrary signal set to a lowpass filter can result in a significant attenuation of higher frequency components, thereby defining a stopband whose boundary is influenced by the choice of filter cutoff frequency, with the unattenuated frequency components defining the filter passband. For instrumentation purposes, approximating the ideal lowpass filter amplitude $A(f)$ and phase $B(f)$ responses described by Figure 3-1 is beneficial in order to achieve signal band-limiting without alteration or the addition of errors to a passband signal of interest. In fact, preserving the accuracy of measurement signals is of sufficient importance that consideration of filter characterizations that correspond to well-behaved functions such as Butterworth and Bessel polynomials are especially useful. However, an ideal filter is physically unrealizable because practical filters are represented by ratios of polynomials that cannot possess the discontinuities required for sharply defined filter boundaries.

Figure 3-2 describes the Butterworth lowpass amplitude response $A(f)$ and Figure 3-3 its phase response $B(f)$, where n denotes the filter order or number of poles. Butterworth filters are characterized by a maximally flat amplitude response in the vicinity of dc, which extends toward its –3 dB cutoff frequency f_c as n increases. This characteristic is defined by equations (3-1) and (3-2) and Table 3-1. Butterworth attenuation is rapid beyond f_c as filter order increases with a slightly nonlinear phase response that provides a good approximation to an ideal lowpass filter. An analysis of the error attributable to this approximation is derived in Section 3-3. Figure 3-4 presents the Butterworth highpass response.

$$B(s) = \left(j\,\frac{f}{f_c}\right)^n + b_{n-1}\left(j\,\frac{f}{f_c}\right)^{n-1} + \cdots + b_0 \qquad (3-1)$$

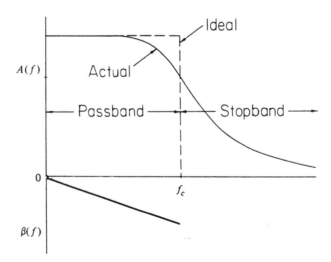

FIGURE 3-1. Ideal lowpass filter.

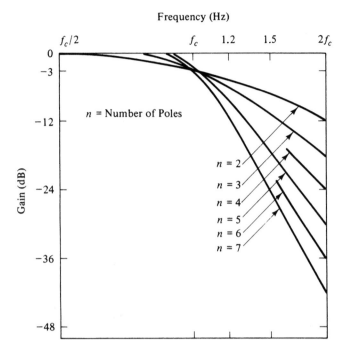

FIGURE 3-2. Butterworth lowpass amplitude.

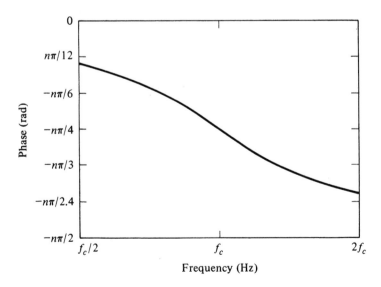

FIGURE 3-3. Butterworth lowpass phase.

TABLE 3-1. Butterworth Polynomial Coefficients

Poles n	b_0	b_1	b_2	b_3	b_4	b_5
1	1.0					
2	1.0	1.414				
3	1.0	2.0	2.0			
4	1.0	2.613	3.414	2.613		
5	1.0	3.236	5.236	5.236	3.236	
6	1.0	3.864	7.464	9.141	7.464	3.864

$$A(f) = \frac{b_0}{\sqrt{B(s)B(-s)}} \tag{3-2}$$

$$= \frac{1}{\sqrt{1 + (f/f_c)^{2n}}}$$

Bessel filters are all-pole filters, like their Butterworth counterparts. with an amplitude response described by equations (3-3) and (3-4) and Table 3-2. Bessel lowpass filters are characterized by a more linear phase delay extending to their cutoff frequency f_c and beyond as a function of filter order n shown in Figure 3-5. However, this linear-phase property applies only to lowpass filters. Unlike the

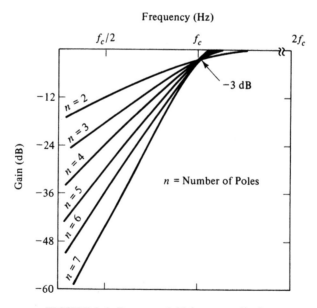

FIGURE 3-4. Butterworth highpass amplitude.

TABLE 3-2. Bessel Polynomial Coefficients

Poles n	b_0	b_1	b_2	b_3	b_4	b_5
1	1					
2	3	3				
3	15	15	6			
4	105	105	45	10		
5	945	945	420	105	15	
6	10,395	10,395	4725	1260	210	21

flat passband of Butterworth lowpass filters, the Bessel passband has no value that does not exhibit amplitude attenuation with a Gaussian amplitude response described by Figure 3-6. It is also useful to compare the overshoot of Bessel and Butterworth filters in Table 3-3, which reveals the Bessel to be much better behaved for bandlimiting pulse-type instrumentation signals and where phase linearity is essential.

$$A(f) = \frac{b_0}{\sqrt{B(s)B(-s)}} \qquad (3\text{-}3)$$

$$B(s) = \left(j\,\frac{f}{f_c}\right)^n + b_{n-1}\left(j\,\frac{f}{f_c}\right)^{n-1} + \cdots + b_0 \qquad (3\text{-}4)$$

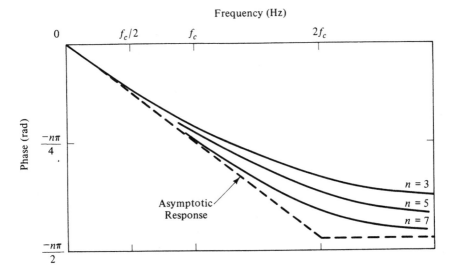

FIGURE 3-5. Bessel lowpass phase.

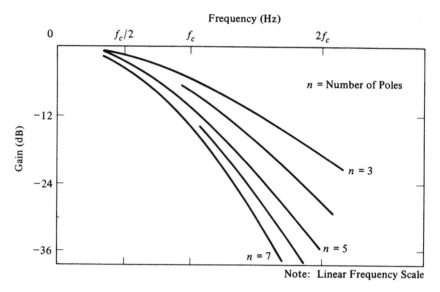

FIGURE 3-6. Bessel lowpass amplitude.

3-2   ACTIVE FILTER NETWORKS

In 1955, Sallen and Key of MIT published a description of 18 active filter networks for the realization of various filter approximations. However, a rigorous sensitivity analysis by Geffe and others disclosed by 1967 that only four of the original networks exhibited low sensitivity to component drift. Of these, the unity-gain and multiple-feedback networks are of particular value for implementing lowpass and bandpass filters, respectively, to Q values of 10. Work by others resulted in the low-sensitivity biquad resonator, which can provide stable Q values to 200, and the stable gyrator band-reject filter. These four networks are shown in Figure 3-7 with key sensitivity parameters. The sensitivity of a network can be determined, for example, when the change in its Q for a change in its passive-element values is evaluated. Equation (3-5) describes the change in the Q of a network by multiplying the thermal coefficient of the component of interest by its sensitivity coefficient. Normally, 50 to 100 ppm/°C components yield good performance.

TABLE 3-3. Filter Overshoot Pulse Response

n	Bessel (%FS)	Butterworth (%FS)
1	0	0
2	0.4	4
3	0.7	8
4	0.8	11

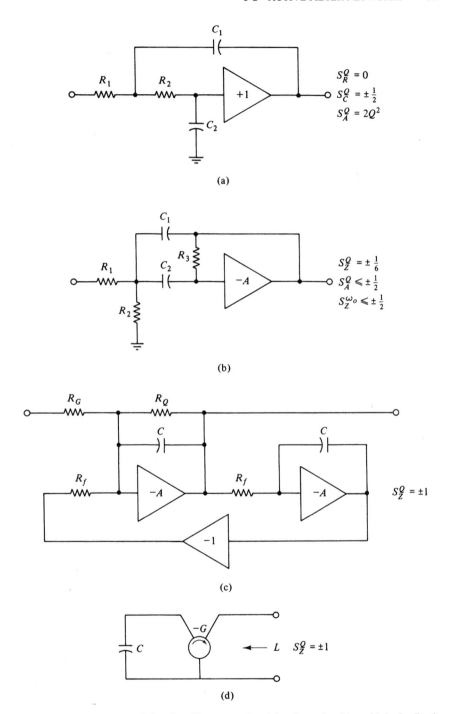

FIGURE 3-7. Recommended active filter networks: (a) unity gain, (b) multiple feedback, (c) biquad, and (d) gyrator.

$$S_z^Q = \pm 1 \text{ passive network} \tag{3-5}$$
$$= (\pm 1)(50 \text{ ppm/}°C)(100\%)$$
$$= \pm 0.005\%Q/°C$$

Unity-gain networks offer excellent performance for lowpass and highpass realizations and may be cascaded for higher-order filters. This is perhaps the most widely applied active filter circuit. Note that its sensitivity coefficients are less than unity for its passive components—the sensitivity of conventional passive networks—and that its resistor temperature coefficients are zero. However, it is sensitive to filter gain, indicating that designs that also obtain greater than unity gain with this filter network are suboptimum. The advantage of the multiple-feedback network is that a bandpass filter can be formed with a single operational amplifier, although the biquad network must be used for high Q bandpass filters. However, the stability of the biquad at higher Q values depends upon the availability of adequate amplifier loop gain at the filter center frequency. Both bandpass networks can be stagger-tuned for a maximally flat passband response when required. The principle of operation of the gyrator is that a conductance $-G$ gyrates a capacitive current to an effective inductive current. Frequency stability is very good, and a band-reject filter notch depth to about -40 dB is generally available. It should be appreciated that the principal capability of the active filter network is to synthesize a complex–conjugate pole pair. This achievement, as described below, permits the realization of any mathematically definable filter approximation.

Kirchoff's current law provides that the sum of the currents into any node is zero. A nodal analysis of the unity-gain lowpass network yields equations (3-6) through (3-9). It includes the assumption that current in C_2 is equal to current in R_2; the realization of this requires the use of a low-input-bias-current operational amplifier for accurate performance. The transfer function is obtained upon substituting for V_x in equation (3-6) its independent expression obtained from equation (3-7). Filter pole positions are defined by equation (3-9). Figure 3-8 shows these nodal equations and the complex-plane pole positions mathematically described by equation (3-9). This second-order network has two denominator roots (two poles) and is sometimes referred to as a resonator.

$$\frac{V_i - V_x}{R_1} = \frac{V_x - V_0}{1/j\omega C_1} + \frac{V_x - V_0}{R_2} \tag{3-6}$$

$$\frac{V_x - V_0}{R_2} = \frac{V_0}{1/j\omega C_2} \tag{3-7}$$

Rearranging,

$$V_x = V_0 \cdot \frac{R_2 + 1/j\omega C_2}{1/j\omega C_2}$$

$$\frac{V_0}{V_i} = \frac{1}{\omega^2 R_1 R_2 C_1 C_2 + \omega C_2 (R_1 + R_2) + 1} \tag{3-8}$$

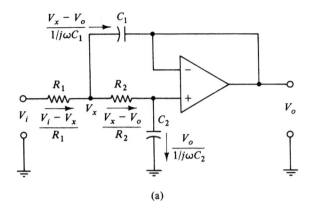

(a)

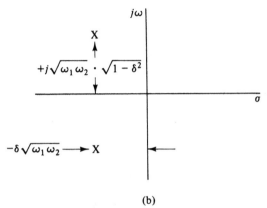

(b)

FIGURE 3-8. Unity-gain network nodal analysis.

$$\omega_1 = \frac{1}{R_1 C_1} \quad \text{and} \quad \omega_2 = \frac{1}{R_2 C_2} \quad \text{and} \quad \delta = \frac{C_2}{2}(R_1 + R_2)$$

$$s_{1,2} = -\delta\sqrt{\omega_1 \omega_2} \pm j\sqrt{\omega_1 \omega_2} \cdot \sqrt{1 - \delta^2} \tag{3-9}$$

A recent technique using MOS technology has made possible the realization of multipole unity-gain network active filters in total integrated circuit form without the requirement for external components. Small-value MOS capacitors are utilized with MOS switches in a switched-capacitor circuit for simulating large-value resistors under control of a multiphase clock. With reference to Figure 3-9 the rate f_s at which the capacitor is toggled determines its charging to V and discharging to V'. Consequently, the average current flow I from V to V' defines an equivalent resistor R that would provide the same average current shown by the identity of

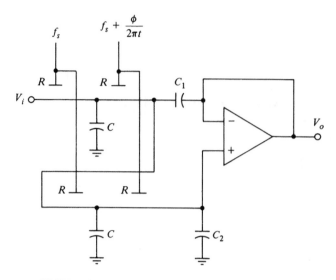

FIGURE 3-9. Switched capacitor unity-gain network.

equation (3-10). The switching rate f_s is normally much higher than the signal frequencies of interest so that the time sampling of the signal can be ignored in a simplified analysis. Filter accuracy is primarily determined by the stability of the frequency of f_s and the accuracy of implementation of the monolithic MOS capacitor ratios.

$$R = \frac{V - V'}{I} = 1/Cf_c \tag{3-10}$$

The most important parameter in the selection of operational amplifiers for active filter service is open-loop gain. The ratio of open-loop to closed-loop gain, or loop gain, must be 10^2 or greater for stable and well-behaved performance at the highest signal frequencies present. This is critical in the application of bandpass filters to ensure a realization that accurately follows the design calculations. Amplifier input and output impedances are normally sufficiently close to the ideal infinite input and zero output values to be inconsequential for impedances in active filter networks. Metal film resistors having a temperature coefficient of 50 ppm/°C are recommended for active filter design.

Selection of capacitor type is the most difficult decision because of many interacting factors. For most applications, polystyrene capacitors are recommended because of their reliable −120 ppm/°C temperature coefficient and 0.05% capacitance retrace deviation with temperature cycling. Where capacitance values above 0.1 μF are required, however, polycarbonate capacitors are available in values to 1 μF with a ±50 ppm/°C temperature coefficient and 0.25% retrace. Mica capacitors are the

most stable devices with ± 50 ppm/°C tempco and 0.1% retrace, but practical capacitance availability is typically only 100 pF to 5000 pF. Mylar capacitors are
available in values to 10 μF with 0.3% retrace, but their tempco averages 400
ppm/°C.

The choice of resistor and capacitor tolerance determines the accuracy of the
filter implementation such as its cutoff frequency and passband flatness. Cost considerations normally dictate the choice of 1% tolerance resistors and 2–5% tolerance capacitors. However, it is usual practice to pair larger and smaller capacitor
values to achieve required filter network values to within 1%, which results in filter parameters accurate to 1 or 2% with low tempco and retrace components.
Filter response is typically displaced inversely to passive-component tolerance,
such as lowering of cutoff frequency for component values on the high side of
their tolerance band. For more critical realizations, such as high-Q bandpass filters, some provision for adjustment provides flexibility needed for an accurate implementation.

Table 3-4 provides the capacitor values in farads for unity-gain networks tabulated according to the number of filter poles. Higher-order filters are formed by a cascade of the second- and third-order networks shown in Figure 3-10, each of which
is different. For example, a sixth-order filter will have six different capacitor values
and not consist of a cascade of identical two-pole or three-pole networks. Figures
3-11 and 3-12 illustrate the design procedure with 1 kHz cutoff, two-pole Butterworth lowpass and highpass filters including the frequency and impedance scaling
steps. The three-pole filter design procedure is identical with observation of the ap-

TABLE 3-4. Unity-Gain Network Capacitor Values in Farads

Poles	Butterworth			Bessel		
	C_1	C_2	C_3	C_1	C_2	C_3
2	1.414	0.707		0.907	0.680	
3	3.546	1.392	0.202	1.423	0.988	0.254
4	1.082	0.924		0.735	0.675	
	2.613	0.383		1.012	0.390	
5	1.753	1.354	0.421	1.009	0.871	0.309
	3.235	0.309		1.041	0.310	
6	1.035	0.966		0.635	0.610	
	1.414	0.707		0.723	0.484	
	3.863	0.259		1.073	0.256	
7	1.531	1.336	0.488	0.853	0.779	0.303
	1.604	0.624		0.725	0.415	
	4.493	0.223		1.098	0.216	
8	1.091	0.981		0.567	0.554	
	1.202	0.831		0.609	0.486	
	1.800	0.556		0.726	0.359	
	5.125	0.195		1.116	0.186	

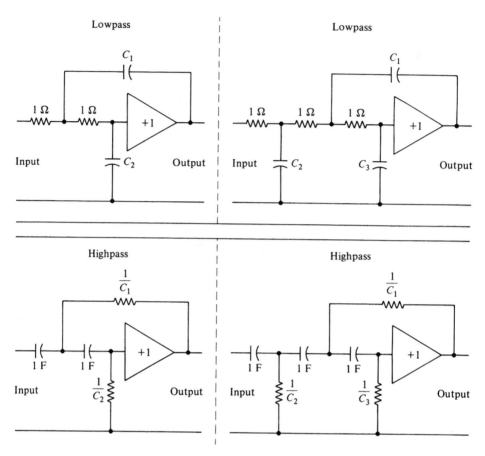

FIGURE 3-10. Two- and three-pole unity-gain networks.

propriate network capacitor locations, but should be driven from a low driving-point impedance such as an operational amplifier. A design guide for unity-gain active filters is summarized in the following steps:

1. Select an appropriate filter approximation and number of poles required to provide the necessary response from the curves of Figures 3-2 through 3-6.
2. Choose the filter network appropriate for the required realization from Figure 3-10 and perform the necessary component frequency and impedance scaling.
3. Implement the filter components by selecting 1% standard-value resistors and then pairing a larger and smaller capacitor to realize each capacitor value to within 1%.

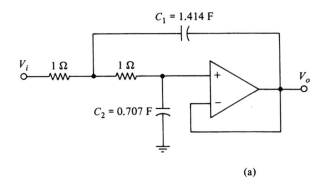

Component values from Table
3-4 are normalized to
1 rad/s with resistors taken
at 1 Ω and capacitors in farads.

(a)

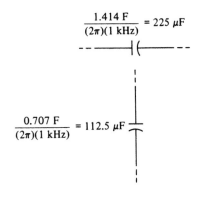

The filter is then
frequency-scaled by
dividing the capacitor
values from the table
by the cutoff frequency
in radians ($2\pi \times 1$ kHz).

(b)

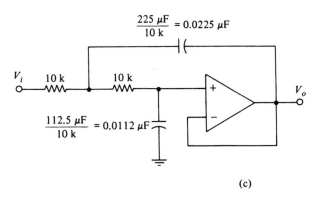

The filter is finally
impedance-scaled by
multiplying the resistor
values by a convenient
value (10 k) and dividing
the capacitor values by
the same value.

(c)

FIGURE 3-11. Butterworth unity-gain lowpass filter example.

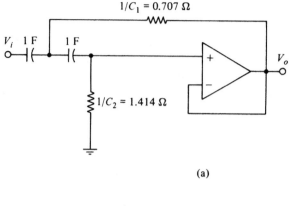

Component values from
Table 3-4 are normalized
to 1 rad/s with capacitors
taken at 1 F and resistors
the inverse capacitor values
from the table in ohms.

(a)

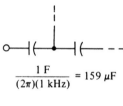

The filter is then frequency-
scaled by dividing the
capacitor values by the
cutoff frequency in radians
of value $(2\pi \times 1 \text{ kHz})$.

$$\frac{1 \text{ F}}{(2\pi)(1 \text{ kHz})} = 159 \ \mu\text{F}$$

(b)

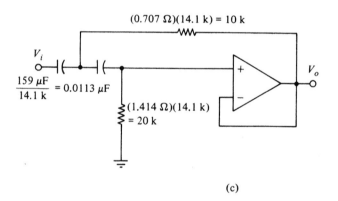

The filter is finally
impedance-scaled by
multiplying the resistor
values by a convenient
value (14.1 k) and
dividing the capacitor
values by the same
value.

(c)

FIGURE 3-12. Butterworth unity-gain highpass filter example.

3-3   FILTER ERROR ANALYSIS

Requirements for signal band-limiting in data acquisition and conversion systems
include signal quality upgrading by signal conditioning circuits, aliasing prevention
associated with sampled-data operations, and intersample error smoothing in output
signal reconstruction. The accuracy, stability, and efficiency of lowpass active filter
networks satisfy most of these requirements with the realization of filter character-

istics appropriate for specific applications. However, when a filter is superimposed on a signal of interest, filter gain and phase deviations from the ideal result in a signal amplitude error that constitutes component error. It is therefore useful to evaluate filter parameters in order to select filter functions appropriate for signals of interest. It will be shown that applying this approach results in a nominal filter error added to the total system error budget. Since dc, sinusoidal, and harmonic signals are encountered in practice, analysis is performed for these signal types to identify optimum filter parameters for achieving minimum error.

Both dc and sinusoidal signals exhibit a single spectral term. Filter gain error is thus the primary source of error because single line spectra are unaffected by filter phase nonlinearities. Figure 3-13 describes the passband gain deviation, with reference to 0 Hz and expressed as average percent error of full scale, for three lowpass filters. The filter error attributable to gain deviation $[1.0 - A(f)]$ is shown to be min-

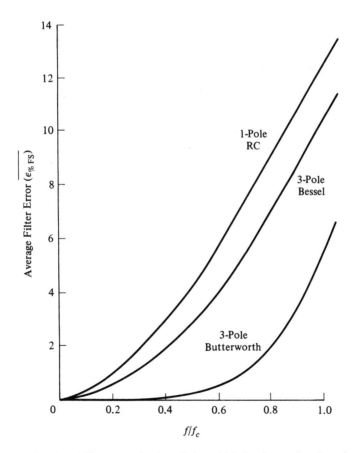

FIGURE 3-13. Plot of filter errors for dc and sinusoidal signals as a function of passband.

imum for the Butterworth characteristic, which is an expected result considering the passband flatness provided by Butterworth filters. Of significance is that small filter component errors can be achieved by restricting signal spectral occupancy to a fraction of the filter cutoff frequency.

Table 3-5 presents a tabulation of the example filters evaluated with dc and sinusoidal signals defining mean amplitude errors for signal bandwidth occupancy to specified filter passband fractions of the cutoff frequency f_c. Equation (3-11) provides an approximate mean error evaluation for RC, Bessel, and Butterworth filter characteristics. Most applications are better served by the three-pole Butterworth filter, which offers a component error of $\overline{0.1}$%FS for signal passband occupancy to 40% of the filter cutoff, plus good stopband attenuation. While it may appear inefficient not to utilize a filter passband up to its cutoff frequency, the total bandwidth sacrificed is usually small. Higher filter orders may also be evaluated when greater stopband attenuation is of interest, with substitution of their amplitude response $A(f)$ in equation (3-11).

$$\overline{\varepsilon_{\%FS}} = \frac{0.1}{\mathrm{BW}/f_c} \sum_{0}^{\mathrm{BW}/f_c} [1.0 - A(f)] \cdot 100\% \qquad \text{(dc and sinusoidal signals)} \qquad (3\text{-}11)$$

The consequence of nonlinear phase delay with harmonic signals is described by Figure 3-14. The application of a harmonic signal just within the passband of a six-pole Butterworth filter provides the distorted output waveform shown. The variation in time delay between signal components at their specific frequencies results in a signal time displacement and the amplitude alteration described. This time variation is apparent from evaluation of equation (3-12), where linear phase provides a constant time delay. A comprehensive method for evaluating passband filter error

TABLE 3-5. Filter Amplitude Errors for dc and Sinusoidal Signals

Signal Bandwidth Passband Fractional Occupancy	Amplitude Response $A(f)$			Average Filter Error $\overline{\varepsilon_{\%FS}}$		
$\dfrac{\mathrm{BW}}{f_c}$	One-Pole RC	Three-Pole Bessel	Three-Pole Butterworth	One-Pole RC	Three-Pole Bessel	Three-Pole Butterworth
0.05	0.999	0.999	1.000	0.1	0.1	0
0.1	0.997	0.998	1.000	0.3	0.2	0
0.2	0.985	0.988	1.000	0.9	0.7	0
0.3	0.958	0.972	1.000	1.9	1.4	0
0.4	0.928	0.951	0.998	3.3	2.3	0.1
0.5	0.894	0.924	0.992	4.7	3.3	0.2
0.6	0.857	0.891	0.977	6.3	4.6	0.7
0.7	0.819	0.852	0.946	8.0	6.0	1.4
0.8	0.781	0.808	0.890	9.7	7.7	2.6
0.9	0.743	0.760	0.808	11.5	9.5	4.4
1.0	0.707	0.707	0.707	13.3	11.1	6.9

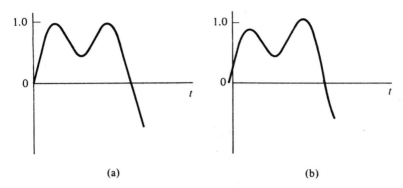

(a)   (b)

FIGURE 3-14. Filtered complex waveform phase nonlinearity. (a) Sum of fundamental and third harmonic in 2:1 ratio. (b) Sum of fundamental and third harmonic following six-pole lowpass Butterworth filter with signal spectral occupancy to filter cutoff.

for harmonic signals is reported by Brockman [14]. An error signal $\varepsilon(t)$ is derived as the difference between the output $y(t)$ of a filter of interest and a delayed input signal $x_0(t)$, expressed by equations (3-13) through (3-15) and described in Figure 3-15. A volts-squared output error is then obtained from the Fourier transform of this error signal and the application of trigonometric identities, and expressed in terms of mean squared error (MSE) by equation (3-16), with A_n and ϕ_n, the filter magnitude and phase responses at n frequencies.

$$\text{Delay variation} = \frac{\phi_a}{2\pi f_a} - \frac{\phi_b}{2\pi f_b} \text{ sec} \tag{3-12}$$

$$y(t) = \sum_{n=1}^{N} A_n \cos(\omega_n t - \phi_n) \tag{3-13}$$

$$x_0(t) = \sum_{n=1}^{N} \cos(\omega_n t - \omega_n t_0) \tag{3-14}$$

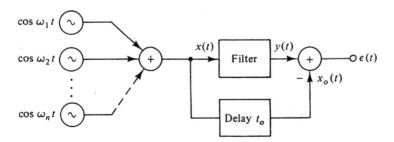

FIGURE 3-15. Filter harmonic signal error analysis.

$$\varepsilon(t) = y(t) - x_0(t) \tag{3-15}$$

$$= \sum_{n=1}^{N} [A_n \cos(\omega_n t - \phi_n) - \cos(\omega_n t - \omega_n t_0)]$$

$$\text{MSE} = \frac{1}{2} \sum_{n=1}^{N} [(A_n \cos \phi_n - \cos \omega_n t_0)^2 + (A_n \sin \phi_n - \sin \omega_n t_0)^2] \tag{3-16}$$

Computer simulation of first through eighth-order Butterworth and Bessel low-pass filters were obtained with the structure of Figure 3-15. The signal delay t_0 was varied in a search for the minimum true MSE by applying the Newton–Raphson method to the derivative of the MSE expression. This exercise was repeated for each filter with various passband spectral occupancies ranging from 10 to 100% of the cutoff frequency, and $N = 10$ sinusoids per octave represented as the simulated input signal $x(t)$. MSE is calculated by the substitution of each t_0 value in equation (3-16), and expressed as average filter component error $\overline{\varepsilon_{\%FS}}$ by equation (3-17) over the filter passband fraction specified for signal occupancy.

$$\overline{\varepsilon_{\%FS}} = \frac{\sqrt{\text{MSE}}}{x(t)} \cdot 100\% \qquad \text{(harmonic signals)} \tag{3-17}$$

Table 3-6 provides a tabulation of these results describing an efficient filter-cut-off-to-signal-bandwidth ratio f_c/BW of 3, considering filter passband signal occupancy versus minimized component error. Signal spectral occupancy up to the filter cutoff frequency is also simulated for error reference purposes. The application of

TABLE 3-6. Filter Amplitude Errors for Harmonic Signals

	Filter Order (Poles)		Average Filter Error $\overline{\varepsilon_{\%FS}}$		
RC	Butterworth	Bessel	$f_c = 10\ \text{BW}$	$f_c = 3\ \text{BW}$	$f_c = \text{BW}$
1			1.201%		
	2			1.093	6.834
		2		0.688	6.179
	3			0.115	5.287
		3		0.677	6.045
	4			0.119	5.947
		4		0.698	6.075
	5			0.134	6.897
		5		0.714	6.118
	6			0.153	7.900
		6		0.725	6.151
	7			0.172	8.943
		7		0.997	6.378
	8			0.195	9.996
		8		1.023	6.299

higher-order filters is primarily determined by the need for increased stopband attenuation compared with the additional complexity and component precision required for their realization.

Lowpass band-limiting filters are frequently required by signal conditioning channels, as illustrated in the following chapters, and especially for presampling antialiasing purposes plus output signal interpolation in sampled-data systems. Of interest is whether the intelligence represented by a signal is encoded in its amplitude values, phase relationships, or combined. Filter mean nonlinearity errors presented in Tables 3-5 and 3-6 describe amplitude deviations of filtered signals resulting from nonideal filter magnitude and phase characteristics. It is clear from these tabulations that Butterworth filters contribute nominal error to signal amplitudes when their passband cutoff frequency is derated to multiples of a signal BW value. It is also notable that measurands and encoded data are so commonly represented by signal amplitude values in instrumentation systems that Butterworth filters are predominant.

When signal phase accuracy is essential for phase-coherent applications, ranging from communications to audio systems, including matrixed home theater signals, then Bessel lowpass filters are advantageous. For example, if only signal phase is of interest, an examination of Figure 3-5 and Tables 3-5 and 3-6 reveal that derating a three-pole Bessel filter passband cutoff frequency to three times the signal BW achieves very linear phase, but signal amplitude error approaches 1%FS. However, error down to 0.1–0.2%FS in both amplitude and phase are provided for any signal type when this lowpass filter is derated on the order of ten times signal BW. At that operating point, Bessel filters behave as pure delay lines to the signal.

3-4   BANDPASS INSTRUMENTATION FILTERS

The bandpass filter passes a band of frequencies of bandwidth Δf centered at a frequency f_0 and attenuates all other frequencies. The quality factor Q of this filter is a measure of its selectivity and is defined by the ratio $f_0/\Delta f$. Also of interest is the geometric mean of the upper and lower -3 dB frequencies defining Δf, or $f_g = f_u \cdot f_L$. Equations (3-18) and (3-19) present the amplitude function for a second-order bandpass filter in terms of these quantities, with amplitude response for various Q values plotted in Figure 3-16.

$$A(f) = \frac{2\pi f_0/Q}{\sqrt{B(s)B(-s)}} \tag{3-18}$$

$$B(s) = (j2\pi f) + \frac{2\pi f_0}{Q} + \frac{(2\pi f_0)^2}{j2\pi f} \tag{3-19}$$

It may be appreciated from this figure that for all Q values the bandpass skirt attenuation rolloff relaxes to -12 dB/octave one octave above and below f_0, which is expected for any second-order filter. (An octave is the interval between two frequencies, one twice the other.) Greater skirt attenuation can be obtained by cascad-

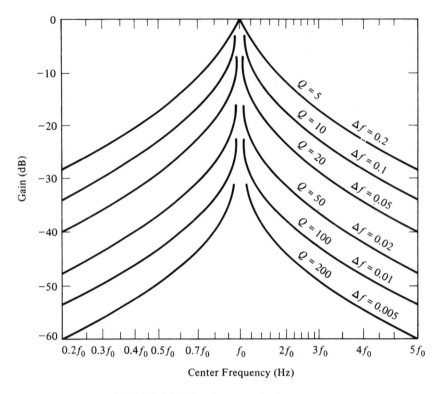

FIGURE 3-16. Bandpass amplitude response.

ing these single-tuned sections, thereby producing a higher-order filter. The phase response of a bandpass filter may be envisioned as that of a highpass and lowpass filter in cascade. This phase has a slope whose change is monotonic and of value $0°$ at f_0, asymptotically reaching its maximum positive and maximum negative phase shift below and above f_0, respectively; total phase shift is a function of the filter order $n \cdot 90°$.

The band-reject filter, also called a band-elimination or notch filter, passes all frequencies except those centered about f_c. Its amplitude function is described by equations (3-20) to (3-22), and its amplitude response by Figure 3-17. Band-reject Q is determined by the ratio $f_c/\Delta f$, where bandwidth Δf is defined between the -3 dB passband cutoff frequencies. Band-reject filter phase response follows the same phase characteristics described for the bandpass filter. For instrumentation service, the band-reject response can be obtained from the lowpass Butterworth coefficients of Table 3-1, and a maximally flat passband can be realized with paralleled Butterworth lowpass plus highpass filters.

$$A(f) = \frac{1}{\sqrt{B(s)B(-s)}} \tag{3-20}$$

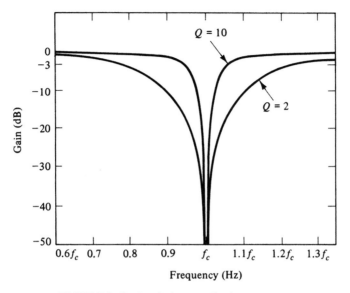

FIGURE 3-17. Bandreject amplitude response.

$$B(s) = C^n + b_{n-1}C^{n-1} + \ldots b_0 \tag{3-21}$$

$$C = \frac{\Delta f(j2\pi f)}{(j2\pi f)^2 + (2\pi f_c)^2} \tag{3-22}$$

A multiple-feedback bandpass filter (MFBF) is shown in Figure 3-18 with a center frequency of 1 Hz and a Q of 10. Equations (3-23) to (3-26) derive the component values for this filter. Normally, a standard capacitor value C is chosen in a range that results in reasonable resistor values with components selected to 1% tolerance. It should be noted that this circuit produces a signal inversion. When higher-Q active bandpass filtering is required, the bi-quad network must be considered. Although its mechanization does require three operational amplifiers, the bi-quad provides the capability to independently set filter center frequency f_0, Q, and gain A_{f_0} at the center frequency. A practical design approach is to fix the frequency-determining resistors R_{f_0} shown in Figure 3-21 at a standard value, and then calculate the other component values as presented by the equations in Table 3-8 for representative instrumentation frequencies.

$$k = 2\pi f_0 C \tag{3-23}$$

$$= (6.28)(1 \text{ Hz})(1 \ \mu\text{F})$$

$$= 6.28 \times 10^{-6} \text{ mho}$$

68

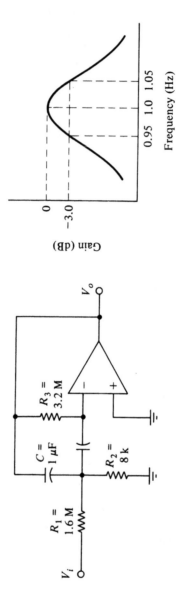

FIGURE 3-18. Multiple-feedback bandpass filter.

$$R_1 = \frac{Q}{k} \tag{3-24}$$

$$= \frac{10}{6.28 \times 10^{-6}}$$

$$= 1.6 \text{ M}$$

$$R_2 = \frac{1}{(2Q - 1/Q)k} \tag{3-25}$$

$$= \frac{1}{(20 - 0.1)(6.28 \times 10^{-6})}$$

$$= 8 \text{ K}$$

$$R_3 = \frac{2Q}{k} \tag{3-26}$$

$$= \frac{20}{6.28 \times 10^{-6}}$$

$$= 3.2 \text{ M}$$

Most instrumentation systems involve amplitude measurements of transducer outputs, and it is normally of interest to maintain amplitude flatness in the signal passband. In the case of bandpass filtering using the previous single-tuned networks, the amplitude response rolls off immediately on both sides of the center frequency. Bandpass signals having an extended spectral occupancy, therefore, should be filtered by a flat-passband bandpass filter. A stagger-tuning scheme for multiple-cascaded, single-tuned bandpass filters can produce a flat passband with the additional benefit of increased skirt selectivity. Table 3-7 presents stagger-

TABLE 3-7. Stagger-Tuning Parameters

Single-Tuned Filters	Δf_r	f_r
2	$0.71 \, \Delta f$	$f_0 + 0.35 \, \Delta f$
	$0.71 \, \Delta f$	$f_0 - 0.35 \, \Delta f$
3	$0.5 \, \Delta f$	$f_0 + 0.43 \, \Delta f$
	$0.5 \, \Delta f$	$f_0 - 0.43 \, \Delta f$
	$1.0 \, \Delta f$	f_0
4	$0.35 \, \Delta f$	$f_0 + 0.46 \, \Delta f$
	$0.38 \, \Delta f$	$f_0 - 0.46 \, \Delta f$
	$0.93 \, \Delta f$	$f_0 + 0.19 \, \Delta f$
	$0.93 \, \Delta f$	$f_0 - 0.19 \, \Delta f$

tuning parameters for a maximally flat passband in terms of the number of single-tuned networks employed, their individual center frequencies f_r and –3 dB bandwidths Δf_r, and the overall bandpass filter center frequency f_0 and –3 dB bandwidth Δf. Passband flatness and skirt selectivity both improve, of course, as the number of cascaded single-tuned networks increases and the overall Δf decreases.

Consider, for example, a bandpass filter requirement centered at an f_0 of 1 kHz with a maximally flat Δf bandwidth of 200 Hz. This $Q = 5$ filter is also to achieve –35 dB attenuation ±1 octave on both sides of the center frequency f_0. Two cascaded and stagger-tuned MFBF networks are able to meet these specifications, requiring only two operational amplifiers for their implementation. The individual MFBF networks are designed according to the example associated with Figure 3-18, but employing the tuning parameters obtained from Table 3-7. The filter circuit is shown by Figure 3-19 with 0.1 µF capacitors. In the event that final minor tuning adjustments are required, each network center frequency is determined by R_2, Q by R_3, and gain by R_1. A penalty of the stagger-tuned method is a gain loss that results from the algebraic addition of the skirts of each network. However, this loss may be calculated and compensated for on a per-network basis as shown in the example calculations that follow. The overall filter response achieved is described by Figure 3-20 and has a +0.3 dB amplitude response at ±70 Hz (70% bandwidth) on either side of the 0 dB f_0.

$$\text{gain loss}_r = \frac{A_r}{\sqrt{A_r^2 + B_r^2}} \tag{3-27}$$

$$A_r = \frac{(2\pi f_r)(2\pi f_g)}{Q_r} \tag{3-28}$$

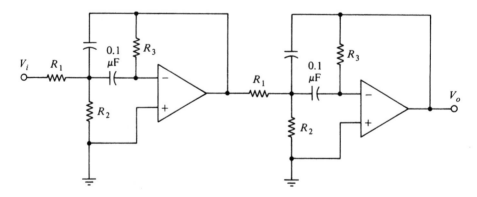

FIGURE 3-19. Stagger-tuned multiple-feedback bandpass filter.

$$B_r = (2\pi f_r)^2 - (2\pi f_g)^2 \tag{3-29}$$

$$f_g = \sqrt{f_u \cdot f_L} \tag{3-30}$$
$$= \sqrt{(1{,}100 \text{ Hz})(900 \text{ Hz})}$$
$$= 995 \text{ Hz}$$

First Section:

$\Delta f_{r_1} = 0.71 \, \Delta f$ (Table 3-7)
$\quad = (0.71)(200 \text{ Hz})$
$\quad = 141 \text{ Hz}$

$f_{r_1} = f_0 + 0.35 \, \Delta f$ (Table 3-7)
$\quad = 1 \text{ kHz} + (0.35)(200 \text{ Hz})$
$\quad = 1.07 \text{ kHz}$

$Q_1 = \dfrac{f_{r_1}}{\Delta f_{r_1}}$ (3-31)

$\quad = \dfrac{1.07 \text{ kHz}}{141 \text{ Hz}}$

$\quad = 7.6$

$k_1 = 2\pi f_{r_1} C$ (3-23)
$\quad = (2\pi)(1.07 \text{ kHz})(0.1 \text{ } \mu\text{F})$
$\quad = 6.72 \times 10^{-4} \text{ mho}$

$A_1 = \dfrac{(2\pi f_{r_1})(2\pi f_g)}{Q_1}$ (3-28)

$\quad = \dfrac{(2\pi \cdot 1.07 \text{ kHz})(2\pi \cdot 995 \text{ Hz})}{7.6}$

$\quad = 5.53 \times 10^6$

$B_1 = (2\pi f_{r_1})^2 - (2\pi f_g)^2$ (3-29)
$\quad = (2\pi \cdot 1.07 \text{ kHz})^2 - (2\pi \cdot 995 \text{ Hz})^2$
$\quad = 6.15 \times 10^6$

Second Section:

$\Delta f_{r_2} = 0.71 \, \Delta f$ (Table 3-7)
$\quad = (0.71)(200 \text{ Hz})$
$\quad = 141 \text{ Hz}$

$f_{r_2} = f_0 - 0.35 \, \Delta f$ (Table 3-7)
$\quad = 1 \text{ kHz} - (0.35)(200 \text{ Hz})$
$\quad = 930 \text{ Hz}$

$Q_2 = \dfrac{f_{r_2}}{\Delta f_{r_2}}$ (3-31)

$\quad = \dfrac{930 \text{ Hz}}{141 \text{ Hz}}$

$\quad = 6.6$

$k_2 = 2\pi f_{r_2} C$ (3-23)
$\quad = (2\pi)(930 \text{ Hz})(0.1 \text{ } \mu\text{F})$
$\quad = 5.85 \times 10^{-4} \text{ mho}$

$A_2 = \dfrac{(2\pi f_{r_2})(2\pi f_g)}{Q_2}$ (3-28)

$\quad = \dfrac{(2\pi \cdot 930 \text{ Hz})(2\pi \cdot 995 \text{ Hz})}{6.6}$

$\quad = 5.53 \times 10^6$

$B_2 = (2\pi f_{r_2})^2 - (2\pi f_g)^2$ (3-29)
$\quad = (2\pi \cdot 930 \text{ kHz})^2 - (2\pi \cdot 995 \text{ Hz})^2$
$\quad = -4.9 \times 10^6$

$$\frac{\text{Gain}}{\text{loss}_1} = \frac{A_1}{\sqrt{A_1^2 + B_1^2}} \quad (3\text{-}27)$$

$$= \frac{5.53 \times 10^6}{\sqrt{(5.53 \times 10^6)^2 + (6.15 \times 10^6)^2}}$$

$$= 0.669$$

$$\frac{\text{Gain}}{\text{loss}_2} = \frac{A_2}{\sqrt{A_2^2 + B_2^2}} \quad (3\text{-}27)$$

$$= \frac{5.53 \times 10^6}{\sqrt{(5.53 \times 10^6)^2 + (-4.9 \times 10^6)^2}}$$

$$= 0.75$$

$$R_1 = \frac{Q_1 \cdot \text{gain loss}_1}{k_1} \quad (3\text{-}32)$$

$$= \frac{(7.6)(0.669)}{6.72 \times 10^{-4} \text{ mho}}$$

$$= 7.55 \text{ K}$$

$$R_1 = \frac{Q_2 \cdot \text{gain loss}_2}{k_2} \quad (3\text{-}32)$$

$$= \frac{(6.6)(0.75)}{5.85 \times 10^{-4} \text{ mho}}$$

$$= 8.47 \text{ K}$$

$$R_2 = \frac{1}{\left(2Q_1 - \dfrac{1}{Q_1 \cdot \text{gain loss}_1}\right)k_1} \quad (3\text{-}33)$$

$$= \frac{1}{\left(15.2 - \dfrac{1}{(7.6)(0.669)}\right)(6.72 \times 10^{-4}) \text{ mho}}$$

$$= 99 \ \Omega$$

$$R_1 = \frac{1}{\left(2Q_2 - \dfrac{1}{Q_2 \cdot \text{gain loss}_2}\right)k_2} \quad (3\text{-}33)$$

$$= \frac{1}{\left(13.2 - \dfrac{1}{(6.6)(0.75)}\right)(5.85 \times 10^{-4}) \text{ mho}}$$

$$= 132 \ \Omega$$

First Section:

$$R_3 = \frac{2Q_1}{k_1} \quad (3\text{-}26)$$

$$= \frac{15.2}{6.72 \times 10^{-4} \text{ mho}}$$

$$= 22.6 \text{ K}$$

Second Section:

$$R_3 = \frac{2Q_2}{k_2} \quad (3\text{-}26)$$

$$= \frac{13.2}{5.85 \times 10^{-4} \text{ mho}}$$

$$= 22.6 \text{ K}$$

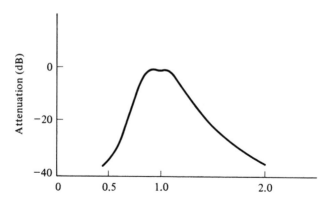

FIGURE 3-20. Stagger-tuned $Q = 5$ bandpass response.

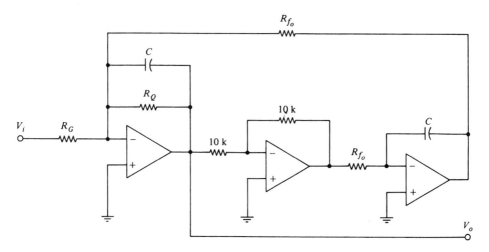

FIGURE 3-21. Biquad bandpass filter.

TABLE 3-8. Biquad Component Values for $R_{f_0} = 10$ K

f_0	$C = \dfrac{1}{(2\pi f_0 R_{f_0})}$	Q	$R_Q = \dfrac{Q}{2\pi f_0 C}$	A_{f_0}	$R_q = \dfrac{R_q}{A_{f_0}}$
10 Hz	1.6 μF	10	100 K	Q/100	1 M
100 Hz	0.16 μF	50	500 K	Q/50	500 K
1 kHz	0.016 μF	100	1 M	Q/10	100 K
10 kHz	0.0016 μF	200	2 M	Q	10 K

The gyrator band-reject filter realization is described by Figure 3-22. Its center-frequency stability is very good, but the realization of Q values greater than about 5 requires an amplifier-loop gain of 10^3 or greater at the notch frequency f_c. Also, at higher Q values, signal-input amplitude may have to be attenuated to preserve linearity. However, a notch depth to −40 dB is available with this circuit.

BIBLIOGRAPHY

1. R. P. Sallen and E. L. Key, "A Practical Method of Designing RC Active Filters," *IRE Transactions on Circuit Theory, CT-2,* March 1955.
2. P. R. Geffe, "Toward High Stability in Active Filters," *IEEE Spectrum, 7,* May 1970.
3. L. C. Thomas, "The Biquad, Part 1—Some Practical Design Considerations," *IEEE Circuit Theory Transactions, CT-18,* May 1971.
4. S. K. Mitra, "Synthesizing Active Filters," *IEEE Spectrum, 6,* January 1969.
5. R. Brandt, "Active Resonators Save Steps in Designing Active Filters," *Electronics* April 24, 1972.
6. B. Zeines, *Introduction to Network Analysis,* Englewood Cliffs, NJ: Prentice Hall, 1967.

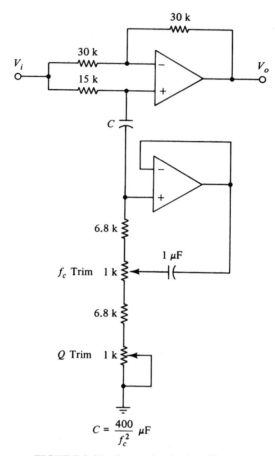

FIGURE 3-22. Gyrator band-reject filter.

7. C. Mitra, *Analysis and Synthesis of Linear Active Networks,* New York, John Wiley, 1969.

8. J. W. Craig, *Design of Lossy Filters,* Cambridge, MA: MIT Press, 1970.

9. R. W. Daniels, *Approximation Methods for Electronic Filter Design,* New York, Mc-Graw-Hill, 1974.

10. D. E. Johnson, *Introduction to Filter Theory,* Englewood Cliffs, NJ: Prentice Hall, 1976.

11. D. E. Johnson and J. L. Hilburn, *Rapid Practical Designs of Active Filters,* New York: Wiley, 1975.

12. P. E. Allen and L. P. Huelsman, *Theory and Design of Active Filters,* New York, Mc-Graw-Hill, 1980.

13. J. D. Rhodes, *Theory of Electrical Filters,* New York: Wiley, 1976.

14. J. P. Brockman, "Interpolation Error in Sampled Data Systems," Electrical Engineering Department, University of Cincinnati, May 1985.

15. S. Laube, "Comparative Analysis of Total Average Filter Component Error," Senior Design Project, Electrical Engineering Technology, University of Cincinnati, 1983.

4

LINEAR SIGNAL CONDITIONING TO SIX-SIGMA CONFIDENCE

4-0 INTRODUCTION

Economic considerations are imposing increased accountability on the design of analog I/O systems to provide performance at the required accuracy for computer-integrated measurement and control instrumentation without the costs of overdesign. Within that context, this chapter provides the development of signal acquisition and conditioning circuits, and derives a unified method for representing and upgrading the quality of instrumentation signals between sensors and data-conversion systems. Low-level signal conditioning is comprehensively developed for both coherent and random interference conditions employing sensor–amplifier–filter structures for signal quality improvement presented in terms of detailed device and system error budgets. Examples for dc, sinusoidal, and harmonic signals are provided, including grounding, shielding, and noise circuit considerations. A final section explores the additional signal quality improvement available by averaging redundant signal conditioning channels, including reliability enhancement. A distinction is made between signal conditioning, which is primarily concerned with operations for improving signal quality, and signal processing operations that assume signal quality already at the level of interest. An overall theme is the optimization of performance through the provision of methods for effective analog design.

4-1 SIGNAL CONDITIONING INPUT CONSIDERATIONS

The designer of high-performance instrumentation systems has the responsibility of defining criteria for determining preferred options from among available alternatives. Figure 4-1 illustrates a cause-and-effect outline of comprehensive methods that are developed in this chapter, whose application aids the realization of effective signal conditioning circuits. In this fishbone chart, grouped system and device op-

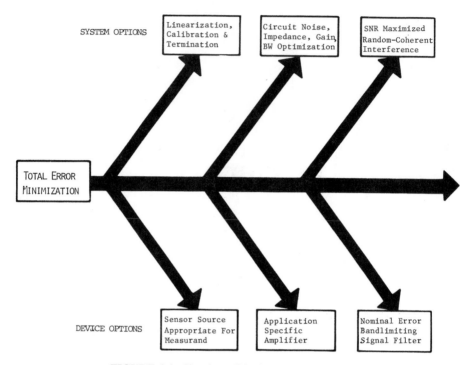

FIGURE 4-1. Signal conditioning design influences.

tions are outlined for contributing to the goal of minimum total instrumentation error. Sensor choices appropriate for measurands of interest were introduced in Chapter 1, including linearization and calibration issues. Application-specific amplifier and filter choices for signal conditioning are defined, respectively, in Chapters 2 and 3. In this section, input circuit noise, impedance, and grounding effects are described for signal conditioning optimization. The following section derives models that combine device and system quantities in the evaluation and improvement of signal quality, expressed as total error, including the influence of random and coherent interference. The remaining sections provide detailed examples of these signal conditioning design methods.

External interference entering low-level instrumentation circuits frequently is substantial and techniques for its attenuation are essential. Noise coupled to signal cables and power buses has as its cause electric and magnetic field sources. For example, signal cables will couple 1 mV of interference per kilowatt of 60 Hz load for each lineal foot of cable run of 1 ft spacing from adjacent power cables. Most interference results from near-field sources, primarily electric fields, whereby an effective attenuation mechanism is reflection by nonmagnetic materials such as copper or aluminum shielding. Both foil and braided shielded twinax signal cable offer attenuation on the order of −90 voltage dB to 60 Hz interference, which degrades by approximately +20 dB per decade of increasing frequency.

For magnetic fields absorption is the effective attenuation mechanism requiring steel or mu metal shielding. Magnetic fields are more difficult to shield than electric fields, where shielding effectiveness for a specific thickness diminishes with decreasing frequency. For example, steel at 60 Hz provides interference attenuation on the order of −30 voltage dB per 100 mils of thickness. Applications requiring magnetic shielding are usually implemented by the installation of signal cables in steel conduit of the necessary wall thickness. Additional magnetic field attenuation is furnished by periodic transposition of twisted-pair signal cable, provided no signal returns are on the shield, where low-capacitance cabling is preferable. Mutual coupling between computer data acquisition system elements, for example from finite ground impedances shared among different circuits, also can be significant, with noise amplitudes equivalent to 50 mV at signal inputs. Such coupling is minimized by separating analog and digital circuit grounds into separate returns to a common low-impedance chassis star-point termination, as illustrated in Figure 4-3.

The goal of shield ground placement in all cases is to provide a barrier between signal cables and external interference from sensors to their amplifier inputs. Signal cable shields also are grounded at a single point, below 1 MHz signal bandwidths, and ideally at the source of greatest interference, where provision of the lowest impedance ground is most beneficial. One instance in which a shield is not grounded is when driven by an amplifier guard. Guarding neutralizes cable-to-shield capacitance imbalance by driving the shield with common-mode interference appearing on the signal leads; this also is known as active shielding.

The components of total input noise may be divided into external contributions associated with the sensor circuit, and internal amplifier noise sources referred to its input. We shall consider the combination of these noise components in the context of band-limited sensor–amplifier signal acquisition circuits. Phenomena associated with the measurement of a quantity frequently involve energy–matter interactions that result in additive noise. Thermal noise V_t is present in all elements containing resistance above absolute zero temperature. Equation (4-1) defines thermal noise voltage proportional to the square root of the product of the source resistance and its temperature. This equation is also known as the Johnson formula, which is typically evaluated at room temperature or 293°K and represented as a voltage generator in series with a noiseless source resistance.

$$V_t = \sqrt{4kTR_s}\,V_{rms}/\sqrt{Hz}$$

$$k = \text{Boltzmann's constant } (1.38 \times 10^{-23} \text{ J/°K}) \qquad (4\text{-}1)$$

$$T = \text{absolute temperature (°K)}$$

$$R_s = \text{source resistance } (\Omega)$$

Thermal noise is not influenced by current flow through its associated resistance. However, a dc current flow in a sensor loop may encounter a barrier at any contact or junction connection that can result in contact noise owing to fluctuating conductivity effects. This noise component has a unique characteristic that varies as the reciprocal of signal frequency $1/f$, but is directly proportional to the value of dc current. The behavior of this fluctuation with respect to a sensor loop source resistance is to produce a contact noise voltage whose magnitude may be estimated at a signal frequency of interest by the empirical relationship of equation (4-2). An important conclusion is that dc current flow should be minimized in the excitation of sensor circuits, especially for low signal frequencies.

$$V_c = (0.57 \times 10^{-9}) R_s \sqrt{\frac{I_{dc}}{f}} \; V_{rms}/\sqrt{Hz} \qquad (4\text{-}2)$$

$$I_{dc} = \text{average dc current (A)}$$

$$f = \text{signal frequency (Hz)}$$

$$R_s = \text{source resistance } (\Omega)$$

Instrumentation amplifier manufacturers use the method of equivalent noise–voltage and noise–current sources applied to one input to represent internal noise sources referred to amplifier input, as illustrated in Figure 4-2. The short-circuit rms input noise voltage V_n is the random disturbance that would appear at the input of a noiseless amplifier, and its increase below 100 Hz is due to internal amplifier $1/f$ contact noise sources. The open circuit rms input noise current I_n similarly arises from internal amplifier noise sources and usually may be disregarded in sensor–amplifier circuits because its generally small magnitude typically results in a negligible input disturbance, except when large source resistances are present. Since all of these input noise contributions are essentially from uncorrelated sources, they are combined as the root-sum-square by equation (4-3). Wide bandwidths and large source resistances, therefore, should be avoided in sensor–amplifier signal acquisition circuits in the interest of noise minimization. Further, additional noise sources encountered in an instrumentation channel following the input gain stage are of diminished consequence because of noise amplification provided by the input stage.

$$V_{NPP} = 6.6 \, [(V_t^2 + V_c^2 + V_n^2)(f_{hi})]^{1/2} \qquad (4\text{-}3)$$

4-2 SIGNAL QUALITY EVALUATION AND IMPROVEMENT

The acquisition of a low-level analog signal that represents some measurand, as in Table 4-2, in the presence of appreciable interference is a frequent requirement. Of concern is achieving a signal amplitude measurement A or phase angle ϕ at the accuracy of interest through upgrading the quality of the signal by means of appropriate signal conditioning circuits. Closed-form expressions are available for determining the error of a signal corrupted by random Gaussian noise or coherent sinusoidal interference. These are expressed in terms of signal-to-noise ratios (SNR) by equations (4-4) through (4-9). SNR is a dimensionless ratio of watts of signal to watts of noise, and frequently is expressed as rms signal-to-interference

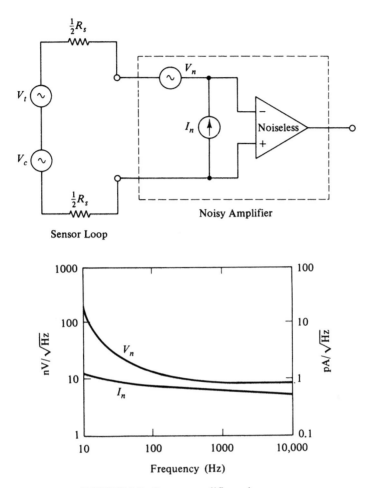

FIGURE 4-2. Sensor–amplifier noise sources.

amplitude squared. These equations are exact for sinusoidal signals, which are typical for excitation encountered with instrumentation sources.

$$P(\Delta A; A) = \text{erf}\left(\frac{1}{2} \frac{\Delta A}{A} \sqrt{\text{SNR}}\right) \text{ probability} \qquad (4\text{-}4)$$

$$0.68 = \text{erf}\left(\frac{1}{2} \frac{\varepsilon_{\%FS}}{100\%} \sqrt{\text{SNR}}\right)$$

$$\varepsilon_{\text{random amplitude}} = \frac{\sqrt{2}\, 100\%}{\sqrt{\text{SNR}}} \text{ of full scale } (1\sigma) \qquad (4\text{-}5)$$

$$P(\Delta\phi; \phi) = \text{erf}\left(\frac{1}{2} \frac{\Delta\phi}{\phi} \sqrt{\text{SNR}}\right) \text{probability} \qquad (4\text{-}6)$$

$$0.68 = \text{erf}\left(\frac{1}{2} \frac{\varepsilon_\phi}{57.3^0/\text{rad}} \sqrt{\text{SNR}}\right)$$

$$\varepsilon_{\text{random phase}} = \frac{1}{2} \cdot \frac{\sqrt{2}\,100}{\sqrt{\text{SNR}}} \text{ degrees } (1\sigma) \qquad (4\text{-}7)$$

$$\varepsilon_{\text{coh amplitude}} = \frac{\Delta A}{A} \cdot 100\% \qquad (4\text{-}8)$$

$$= \sqrt{\frac{V_{\text{coh}}^2}{V_{\text{FS}}^2}} \cdot 100\%$$

$$= \frac{100\%}{\sqrt{\text{SNR}}} \text{ of full scale}$$

$$\varepsilon_{\text{coh phase}} = \frac{100}{2\sqrt{\text{SNR}}} \text{ degrees} \qquad (4\text{-}9)$$

The probability that a signal corrupted by random Gaussian noise is within a specified Δ region centered on its true amplitude A or phase ϕ values is defined by equations (4-4) and (4-6). Table 4-1 presents a tabulation from substitution into these equations for amplitude and phase errors at a 68% (1σ) confidence in their measurement for specific SNR values. One sigma is an acceptable confidence level

TABLE 4-1. SNR Versus Amplitude and Phase Errors

SNR	Amplitude Error Random $\varepsilon_{\%FS}$	Phase Error Random $\varepsilon_{\phi\,deg}$	Amplitude Error Coherent $\varepsilon_{\%FS}$
10^1	44.0	22.3	31.1
10^2	14.0	7.07	9.9
10^3	4.4	2.23	3.1
10^4	1.4	0.707	0.990
10^5	0.44	0.223	0.311
10^6	0.14	0.070	0.099
10^7	0.044	0.022	0.0311
10^8	0.014	0.007	0.0099
10^9	0.0044	0.002	0.0031
10^{10}	0.0014	0.0007	0.00099
10^{11}	0.00044	0.0002	0.00031
10^{12}	0.00014	0.00007	0.00009

TABLE 4-2. Signal Bandwidth Requirements

Signal	Bandwidth (Hz)
dc	$dV_s/\pi V_{FS}dt$
Sinusoidal	1/period T
Harmonic	10/period T
Single event	2/width τ

for many applications. For 95% (2σ) confidence, the error values are doubled for the same SNR. These amplitude and phase errors are closely approximated by the simplifications of equations (4-5) and (4-7), and are more readily evaluated than by equations (4-4) and (4-6). For coherent interference, equations (4-8) and (4-9) approximate amplitude and phase errors where ΔA is directly proportional to V_{coh}, as the true value of A is to V_{FS}. Errors due to coherent interference are seen to be less than those due to random interference by the $\sqrt{2}$ for identical SNR values. Further, the accuracy of these analytical expressions requires minimum SNR values of one or greater. This is usually readily achieved in practice by the associated signal conditioning circuits illustrated in the examples that follow. Ideal matched filter signal conditioning makes use of both amplitude and phase information in upgrading signal quality, and is implied in these SNR relationships for amplitude and phase error in the case of random interference.

For practical applications the SNR requirements ascribed to amplitude and phase error must be mathematically related to conventional amplifier and linear filter signal conditioning circuits. Figure 4-3 describes the basic signal conditioning structure, including a preconditioning amplifier and postconditioning filter and their bandwidths. Earlier work by Fano [1] showed that under high-input SNR conditions, linear filtering approaches matched filtering in its efficiency. Later work by Budai [2] developed a relationship for this efficiency expressed by the characteristic curve of Figure 4-4. This curve and its k parameter appears most reliable for filter numerical input SNR values between about 10 and 100, with an efficiency k of 0.9 for SNR values of 200 and greater.

Equations (4-10) through (4-13) describe the relationships upon which the improvement in signal quality may be determined. Both rms and dc voltage values are interchangeable in equation (4-10). The R_{cm} and R_{diff} impedances of the amplifier input termination account for the V^2/R transducer gain relationship of the input SNR in equation (4-11). CMRR is squared in this equation in order to convert its ratio of differential to common-mode voltage gains to a dimensionally correct power ratio. Equation (4-12) represents the processing–gain relationship for the ratio of amplifier f_{hi} to filter f_c produced with the filter efficiency k, for improving signal quality above that provided by the amplifier CMRR with random interference. Most of the improvement is provided by the amplifier CMRR owing to its squared factor, but random noise higher-frequency components are also effectively attenuated by linear filtering.

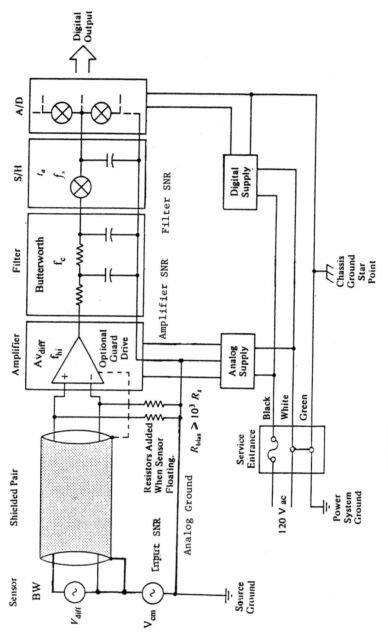

FIGURE 4-3. Signal acquisition system interfaces.

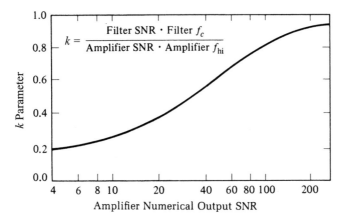

FIGURE 4-4. Linear filter efficiency k versus SNR.

$$\text{Input SNR} = \left(\frac{V_{\text{diff}}}{V_{\text{cm}}}\right)^2 \text{ dc or rms} \tag{4-10}$$

$$\text{Amplifier SNR} = \text{input SNR} \cdot \frac{R_{\text{cm}}}{R_{\text{diff}}} \cdot \text{CMRR}^2 \tag{4-11}$$

$$\text{Filter SNR}_{\text{random}} = \text{amplifier SNR} \cdot k \cdot \frac{f_{\text{hi}}}{f_c} \tag{4-12}$$

$$\text{Filter SNR}_{\text{coherent}} = \text{amplifier SNR} \cdot \left[1 + \left(\frac{f_{\text{coh}}}{f_c}\right)^{2n}\right] \tag{4-13}$$

For coherent interference conditions, signal quality improvement is a function of achievable filter attenuation at the interfering frequency(ies). This is expressed by equation (4-13) for one-pole RC to n-pole Butterworth lowpass filters. Note that filter cutoff frequency is determined from the considerations of Tables 3-5 and 3-6 with regard to minimizing the filter component error contribution. Finally, the various signal conditioning device errors and output signal quality must be appropriately combined in order to determine total channel error. Sensor nonlinearity, amplifier, and filter errors are combined with the root-sum-square of signal errors as described by equation (4-14).

$$\varepsilon_{\text{channel}} = \overline{\varepsilon}_{\text{sensor}} + \overline{\varepsilon}_{\text{filter}} + [\varepsilon_{\text{amplifier}}^2 + \varepsilon_{\text{random}}^2 + \varepsilon_{\text{coherent}}^2]^{1/2} \tag{4-14}$$

Amplitude and phase errors are obtained from the SNR relationships through appropriate substitution in equations (4-4) to (4-9). Substitutions are conveniently provided by equations (4-15) and (4-16), respectively, for coherent and random amplitude error. Observe that these signal quality representations replace the V_{cm}/CMRR entry in Table 2-4 when more comprehensive signal conditioning is employed.

$$\varepsilon_{coherent} = \frac{V_{cm}}{V_{diff}} \cdot \left[\frac{R_{diff}}{R_{cm}} \right]^{1/2} \cdot \frac{A_{V_{cm}}}{A_{V_{diff}}} \cdot \left[1 + \left(\frac{f_{coh}}{f_c} \right)^{2n} \right]^{-1/2} \cdot 100\% \quad (4\text{-}15)$$

$$\varepsilon_{random} = \frac{V_{cm}}{V_{diff}} \cdot \left[\frac{R_{diff}}{R_{cm}} \right]^{1/2} \cdot \frac{A_{V_{cm}}}{A_{V_{diff}}} \cdot \left[\frac{2}{k} \frac{f_c}{f_{hi}} \right]^{1/2} \cdot 100\% \quad (4\text{-}16)$$

4-3   DC, SINUSOIDAL, AND HARMONIC SIGNAL CONDITIONING

Signal conditioning is concerned with upgrading the quality of a signal to the accuracy of interest coincident with signal acquisition, scaling, and band-limiting. The unique requirements of each analog data acquisition channel plus the economic constraint of achieving only the performance necessary in specific applications are an impediment to standardized designs. The purpose of this chapter therefore is to develop a unified, quantitative design approach for signal acquisition and conditioning that offers new understanding and accountability measures. The following examples include both device and system errors in the evaluation of total signal conditioning channel error.

A dc and sinusoidal signal conditioning channel is considered that has widespread industrial application in process control and data logging systems. Temperature measurement employing a Type-C thermocouple is to be implemented over the range of 0 to 1800 °C while attenuating ground conductive and electromagnetically coupled interference. A 1 Hz signal bandwidth (BW) is coordinated with filter cutoff to minimize the error provided by a single-pole filter as described in Table 3-5. Narrowband signal conditioning is accordingly required for the differential-input 17.2 μV/°C thermocouple signal range of 0–31 mV dc, and for rejecting 1 V rms of 60 Hz common mode interference, providing a residual coherent error of 0.009%FS. An OP-07A subtractor instrumentation amplifier circuit combining a 22 Hz differential lag RC lowpass filter is capable of meeting these requirements, including a full-scale output signal of 4.096 V dc with a differential gain $A_{V_{diff}}$ of 132, without the cost of a separate active filter.

This austere dc and sinusoidal circuit is shown by Figure 4-5, with its parameters and defined error performance tabulated in Tables 4-3 through 4-5. This $A_{V_{diff}}$ further results in a –3dB frequency response of 4.5kHz to provide a sensor loop internal noise contribution of 4.4 μV$_{pp}$ with 100 ohms source resistance. With 1% tolerance resistors, the subtractor amplifier presents a common mode gain of 0.02 by the considerations of Table 2-2. The OP-07A error budget of 0.103%FS is combined with other channel error contributions including a mean filter error of 0.1%FS and 0.011%FS linearized thermocouple. The total channel error of 0.246%FS at 1σ expressed in Table 4-5 is dominated by static mean error that is an inflexible error to

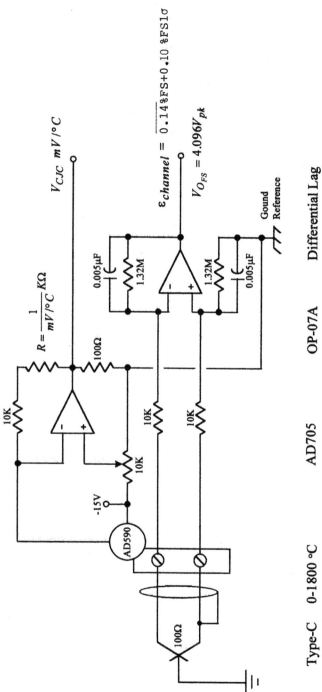

FIGURE 4-5. DC and sinusoidal signal conditioning.

Type-C   0-1800 °C

BW = 1 Hz

V_{diff} = 31mVdc @ 1800 °C

V_{cm} = 1V rms 60Hz coh

AD705

CJC Amplifier

OP-07A

f_{hi} = 4.5KHz

Av_{diff} = 132

Differential Lag

f_c = 22Hz

$V_{CJC}\ mV/°C$

$R = \dfrac{1}{mV/°C}\ K\Omega$

$\varepsilon_{channel} = \overline{0.14\%FS + 0.10\ \%FS1\sigma}$

$V_{O_{FS}}$ = 4.096V_{pk}

Gound
Reference

TABLE 4-3. Amplifier Input Parameters

Symbol	OP-07A	AD624C	AD215BY	Comment
V_{OS}	10 μV	25 μV	0.4 mV	Offset voltage
$\dfrac{dV_{OS}}{dT}$	0.2 μV/°C	0.25 μV/°C	2 μV/°C	Voltage drift
I_{OS}	0.3 nA	10 nA	300 nA	Offset current
$\dfrac{dI_{OS}}{dT}$	5 pA/°C	20 pA/°C	1 nA/°C	Current drift
$A_{V_{\text{diff}}}$	132	50	1	Differential gain
$A_{V_{\text{cm}}}$	0.02 (1%R)	0.0001	0.00001	Common mode gain
CMRR	6600	5×10^5	10^5	$A_{V_{\text{diff}}}/A_{V_{\text{cm}}}$
V_{CM}	10 V_{rms}	10 V_{rms}	1500 V_{rms}	Maximum common mode volts
$V_{N_{\text{pp}}}$	$6.6[(V_t^2 + V_n^2)f_{\text{hi}}]^{1/2}$	$6.6[(V_t^2 + V_c^2 + V_n^2)f_{\text{hi}}]^{1/2}$	$6.6[(V_t^2 f_{\text{hi}})]^{1/2}$	Total input noise
$V_t$rms	1.3 nV/$\sqrt{\text{Hz}}$	4 nV/$\sqrt{\text{Hz}}$	0.9 nV/$\sqrt{\text{Hz}}$	Thermal noise
$V_c$rms	None	1.8 nV/$\sqrt{\text{Hz}}$	Negligible	Contact noise
$V_n$rms	10 nV/$\sqrt{\text{Hz}}$	4 nV/$\sqrt{\text{Hz}}$	Negligible	Amplifier noise
f_{hi}	4.5 KHz	150 KHz	120 KHz	−3db bandwidth
f_{contact}	None	100 Hz	100 Hz	Contact noise frequency
$\dfrac{dA_V}{dT}$	50 ppm/°C	5 ppm/°C	15 ppm/°C	Gain drift
$f(A_V)$	0.01%	0.001%	0.005%	Gain nonlinearity
R_{diff}	$8 \times 10^7\ \Omega$	$10^9\ \Omega$	$10^{12}\ \Omega$	Differential resistance
R_{cm}	$2 \times 10^{11}\ \Omega$	$10^9\ \Omega$	$5 \times 10^9\ \Omega$	Common mode resistance
R_S	100 Ω	1 K	50 Ω	Source resistance
$V_{O_{\text{FS}}}$	4.096 V_{pk}	±5 V_{pp}	±5 V_{pp}	Full-scale output
dT	10°C	10°C	10°C	Temperature variation

be minimized throughout all instrumentation systems. Postconditioning linearization software achieves a residual deviation from true temperature values of 0.2°C over 1800°C, and active cold junction compensation of ambient temperature is provided by an AD590 sensor attached to the input terminal strip to within $\overline{0.5}$°C. Note that R_i is 10 K ohms.

The information content of instrumentation signals is described by their amplitude variation with time or, through Fourier transformation, by signal BW in Hz. Instrumentation signal types are accordingly classified in Table 4-2, with their minimum BW requirements specified in terms of signal waveform parameters. DC signal time rate of change is equated to the time derivative of a sinusoidal signal evaluated at its zero crossing to determine its BW requirement. In the case of harmonic signals, a first-order rolloff of -20dB/decade is assumed from a full-scale signal amplitude at the inverse waveform period $1/T$, defining the fundamental frequency, declining to one-tenth of full scale at a BW value of ten times the fundamental frequency.

Considered now is the premium harmonic signal conditioning channel of Figure 4-6, employing a 0.1%FS systematic error piezoresistive sensor that can transduce acceleration signals in response to applied mechanical force. Postconditioning signal processing options include subsequent signal integration to obtain velocity and then displacement vibration spectra from these acceleration signals by means of an ac integrator, as shown in Figure 2-14, or by digital signal processing. A harmonic signal spectral bandwidth is allowed for this example from dc to 1 KHz with the 1 K source resistance bridge sensor generating a maximum input signal amplitude of 70 mV rms, up to 100 Hz fundamental frequencies, with rolloff at -20db per decade of frequency to 7 mV rms at 1 KHz BW. The ± 0.5 V dc bipolar sensor excitation is furnished by isolated three-terminal regulators to within ± 50 μV dc variation, providing a negligible 0.01%FS differential mode error. The sensor shield buffered common-mode voltage active drive also preserves signal conditioning CMRR over extended cable lengths.

An AD624C preamplifier raises the differential sensor signal to a ± 5V$_{pp}$ full-scale value while attenuating 1 V rms of common mode random interference, in concert with the lowpass filter, to a residual error of 0.006%FS, as defined by equation (4-16). The error budgets of the preamplifier and isolation amplifier, tabulated in Tables 4-3 and 4-4, also include a sensor loop internal noise contribution of 15 μV$_{pp}$ based on the provisions of Figure 4-2, where the $1/f$ contact noise frequency is taken as 10% of signal BW. Three contributions comprising this internal noise are evaluated as source resistance thermal noise V_t, contact noise V_c arising from 1 mA of dc current flow, and amplifier internal noise V_n. The three-pole Butterworth lowpass filter cutoff frequency is derated to a value of 3 BW to minimize its device error. Note that the AD705 filter amplifier is included in the mean filter device error of $\overline{0.115}$%FS. The total channel 1σ instrumentation error of 0.221%FS consists of an approximate equal sum of static mean and variable systematic error values at one-sigma confidence in Table 4-5. Six-sigma confidence is defined by the extended value of 0.75%FS, consisting of one mean plus six RSS error values.

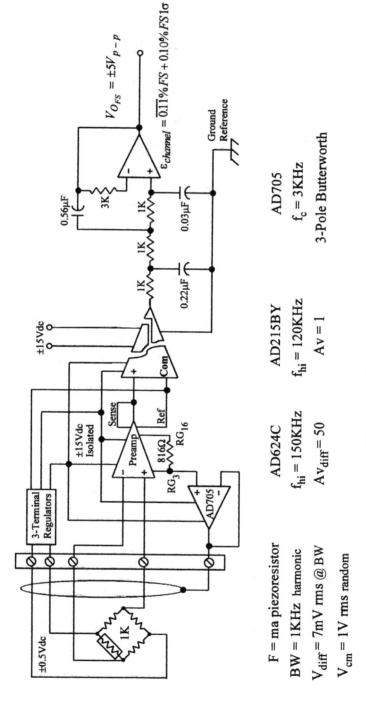

FIGURE 4-6. Premium harmonic signal conditioning.

$VO_{FS} = \pm 5V\, p-p$

$\varepsilon_{channel} = \overline{0.11\% FS + 0.10\% FS1\sigma}$

Ground Reference

$\pm 15Vdc$

0.56µF

3K

1K

1K

1K

0.03µF

0.22µF

Com

Sense

Ref

Preamp

$\pm 15Vdc$ Isolated

816Ω

RG_{16}

RG_3

AD705

3-Terminal Regulators

$\pm 0.5Vdc$

1K

AD624C	AD215BY	AD705
$f_{hi} = 150KHz$	$f_{hi} = 120KHz$	$f_c = 3KHz$
$Av_{diff} = 50$	$Av = 1$	3-Pole Butterworth

F = ma piezoresistor
BW = 1KHz harmonic
V_{diff} = 7mV rms @ BW
V_{cm} = 1V rms random

dc and Sinusoidal Channel	**Harmonic Channel**
Sensor	

dc and Sinusoidal Channel	Harmonic Channel
Type-C thermocouple 17.2 µV/°C post-conditioning linearization	1 KΩ piezoresistor bridge with F = mA response
software $\dfrac{\overline{0.2°C}}{1800°C} \cdot 100\% = \overline{0.011}\%FS$	0.1%FS

Interface	

dc and Sinusoidal Channel	Harmonic Channel
AD 590 temperature sensor cold-junction compensation	Regulators for sensor excitation
$\dfrac{\overline{0.5°C}}{1800°C} \cdot 100\% = 0.032\%FS$	±0.5 V dc ± 50 µV or 0.01%FS

Signal Quality	

dc and Sinusoidal Channel:

$$\varepsilon_{coh} = \frac{V_{cm}}{V_{diff}} \cdot \left[\frac{R_{diff}}{R_{cm}}\right]^{1/2} \cdot \frac{A_{V_{cm}}}{A_{V_{diff}}}$$

$$\cdot \left[1 + \left(\frac{f_{coh}}{f_c}\right)^{2n}\right]^{-1/2} \cdot 100\%$$

$$= \frac{(1\ V_{rms}\ 2\sqrt{2})_{pp}}{31\ mV_{dc}} \cdot \left[\frac{80\ M\Omega}{200\ G\Omega}\right]^{1/2}$$

$$\cdot \frac{0.02}{132} \cdot \left[1 + \left(\frac{60\ Hz}{22\ Hz}\right)^2\right]^{-1/2} \cdot 100\%$$

$$= 0.009\%FS$$

Harmonic Channel:

$$\varepsilon_{rand} = \frac{V_{cm}}{V_{diff}} \cdot \left[\frac{R_{diff}}{R_{cm}}\right]^{1/2} \cdot \frac{A_{V_{cm}}}{A_{V_{diff}}}$$

$$\cdot \left[\frac{2}{k}\frac{f_c}{f_{hi}}\right]^{-1/2} \cdot 100\%$$

$$= \frac{1\ V}{7\ mV} \cdot \left[\frac{1\ G\Omega}{1\ G\Omega}\right]^{1/2} \cdot \frac{10^{-4}}{50}$$

$$\cdot \left[\frac{2}{0.9}\frac{3\ kHz}{150\ kHz}\right]^{1/2} \cdot 100\%$$

$$= 0.006\%FS$$

4-4 REDUNDANT SIGNAL CONDITIONING AND DIAGNOSTICS

When achievable analog signal conditioning error does not meet minimum measurement requirements, identical channels may be averaged to reduce the total error. Random and systematic errors added to the value of a measurement can be reduced by taking the arithmetic mean of a sum of n independent measurement values. This assumes that combined systematic error contributions are sufficient in number to approximate a zero mean value, and as well for random errors. Sensor device error is frequently oversimplified in its specification as the nonlinearity of its transfer function and conservatively represented by a mean error. However, many effects actually contribute to sensor error, such as material–energy interactions,

TABLE 4-4. Amplifier Error Budgets

$\varepsilon_{\text{ampl}_{\text{RTI}}}$	OP07A	AD624C	AD215BY
V_{OS}	$\overline{10}\ \mu\text{V}$	Trimmed	Trimmed
$\dfrac{dV_{\text{OS}}}{dT} \cdot dT$	$2\ \mu\text{V}$	$2.5\ \mu\text{V}$	$20\ \mu\text{V}$
$I_{\text{OS}} \cdot R_i$	$\overline{3}\ \mu\text{V}$	$\overline{10}\ \mu\text{V}$	$\overline{15}\ \mu\text{V}$
$V_{N\text{pp}}$	$4.4\ \mu\text{V}$	$15\ \mu\text{V}$	$2\ \mu\text{V}$
$f(A_V) \cdot \dfrac{V_{O\text{FS}}}{A_{V_{\text{diff}}}}$	$\overline{3}\ \mu\text{V}$	$\overline{1}\ \mu\text{V}$	$\overline{250}\ \mu\text{V}$
$\dfrac{dA_{\text{V}}}{dT} \cdot dT \cdot \dfrac{V_{O\text{FS}}}{A_{V_{\text{diff}}}}$	$15.5\ \mu\text{V}$	$5\ \mu\text{V}$	$750\ \mu\text{V}$
$\Sigma\ \overline{\text{mean}}$ + RSS other	$(\overline{16} + 16)\mu\text{V}$	$(\overline{11} + 16)\mu\text{V}$	$(\overline{265} + 750)\mu\text{V}$
$X\dfrac{A_{V_{\text{diff}}}}{V_{O\text{FS}}} \cdot 100\%$	0.103%FS	0.027%FS	0.020%FS

TABLE 4-5. Signal Conditioning Channel Error Summary

	DC, Sinusoidal		Harmonic	
Element	$\varepsilon_{\%\text{FS}}$	Comment	$\varepsilon_{\%\text{FS}}$	Comment
Sensor	$\overline{0.011}$	Type-C linearized	0.100	Piezoresistor
Interface	$\overline{0.032}$	CJC sensor	0.010	Sensor excitation
Amplifier	0.103	OP-07A	0.027	AD624C
Isolator	None		0.020	AD215AY
Filter	$\overline{0.100}$	Table 3-5	$\overline{0.115}$	Table 3-6
Signal quality	0.009	60 Hz ε_{coh}	0.006	Noise $\varepsilon_{\text{rand}}$
	$\overline{0.143}$%FS	$\Sigma\ \overline{\text{mean}}$	$\overline{0.115}$%FS	$\Sigma\ \overline{\text{mean}}$
$\varepsilon_{\text{channel}}$	0.103%FS	1σ RSS	0.106%FS	1σ RSS
	0.246%FS	$\Sigma\ \overline{\text{mean}} + 1\sigma$ RSS	0.221%FS	$\Sigma\ \overline{\text{mean}} + 1\sigma$ RSS
	0.761%FS	$\Sigma\ \overline{\text{mean}} + 6\sigma$ RSS	0.751%FS	$\Sigma\ \overline{\text{mean}} + 6\sigma$ RSS

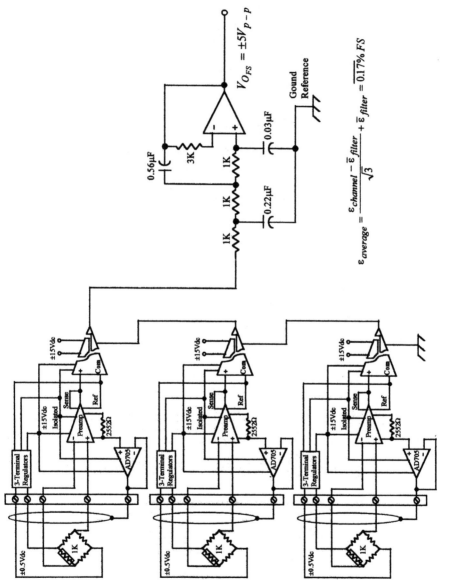

$$\varepsilon_{average} = \frac{\varepsilon_{channel} - \overline{\varepsilon}_{filter}}{\sqrt{3}} + \overline{\varepsilon}_{filter} = \overline{0.17\% \, FS}$$

FIGURE 4-7. Signal conditioning error averaging.

which are unknown, except for their dependence on random variables that generally are compliant to reduction by arithmetic mean averaging.

The foregoing conditions are sufficiently met by typical signal conditioning channels to enable averaged outputs consisting of arithmetic signal additions and RSS error additions. This provides signal quality improvement by $n/\sqrt{n}$ and channel error reduction by its inverse. Averaged measurement error accordingly corresponds to the error of any one identical channel divided by $\sqrt{n}$. However, diminishing returns may result in an economic penalty to achieve error reduction beyond a few channels combined. Further, signal conditioning mean filter device error also remains additive, which is a limitation remedied by relocating the channel filter postaveraging.

Figure 4-7 describes signal conditioning channel averaging in which amplifier stacking between respective device outputs and ground references provide arithmetic signal additions, and their parallel inputs RSS error additions. The $A_{V_{\text{diff}}}$ values of each stage are equally scaled so that the sum of n outputs achieves the full-scale value for a channel within allowable power supply limitations. The three averaged harmonic signal conditioning channels, therefore, each require an $A_{V_{\text{diff}}}$ of 16.67 for a per-channel output of 1.667 volts by employing gain resistors of 2552 ohms. With reference to Table 4-5, moving the filter postaveraging provides an improved overall error of $(0.221\% - \overline{0.115\%})/\sqrt{3} + \overline{0.115\%}$, approximately totaling $\overline{0.176\%}$FS. Note that this connection obviates the requirement for an output summing amplifier and its additional device error contribution.

Redundant structures permit failure detection as well as error averaging for re-

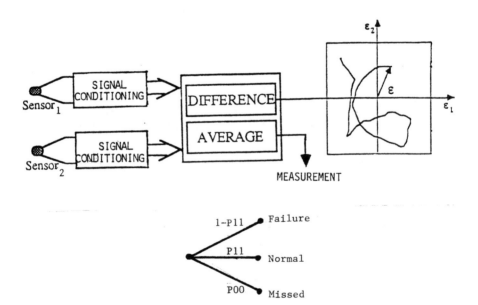

FIGURE 4-8. Dual redundant channel diagnostics.

mote instrumentation diagnostic purposes. The dual redundant signal conditioning channels of Figure 4-8 demonstrate improvement over single channel measurement error with detection of a failure in either channel. Possible system states are illustrated by the decision tree representing their a priori probabilities. Averaging two identical channels provides an expected measurement error improvement of $n/\sqrt{n}$, or 0.707 the error of a single channel. The dual-difference 2D-error vector ε typically exhibits random walk within error space limits defined by the analytically modeled per-channel measurement error exampled in this chapter. Channel output signal amplitude differences exceeding modeled error limits then indicate the failure of a channel. Note that there also exists a finite, but low, probability of missed detection when both channels fail simultaneously with equal output signal amplitudes. Owing to the absence of capability for failed channel isolation with dual redundancy, failure detection is provided without the fail-operational performance that more complex redundant structures, not shown, can allow.

BIBLIOGRAPHY

1. R. M. Fano, "Signal to Noise Ratio in Correlation Detectors," *MIT Technical Report 186,* 1951.

2. M. Budai, "Optimization of the Signal Conditional Channel," Senior Design Project, Electrical Engineering Technology, University of Cincinnati, 1978.

3. H. R. Raemer, *Statistical Communications Theory and Applications,* Englewood Cliffs, NJ: Prentice-Hall, 1969.

4. M. Schwartz, W. Bennett, and S. Stein, *Communications Systems and Techniques,* New York: McGraw-Hill, 1966.

5. P. H. Garrett, *Analog Systems for Microprocessors and Minicomputers,* Reston, VA: Reston, 1978.

6. P. H. Garrett, "Optimize Transducer/Computer Interfaces," *Electronic Design,* May 24, 1977.

7. J. I. Smith, *Modern Operational Amplifier Circuit Design,* New York: Wiley, 1971.

8. J. A. Connelly, *Analog Integrated Circuits,* New York: Wiley-Interscience, 1975.

9. J. M. Pettit and M. M. McWhorter, *Electronic Amplifier Circuits,* New York: McGraw-Hill, 1961.

10. E. M. Petriu (Ed.), Instrumentation and Measurement Technology and Applications, *IEEE Selected Conference Papers,* 1998.

11. B. M. Gordon, *The Analogic Data-Conversion Systems Digest,* Wakefield, MA: Analogic, 1977.

12. *Designers Reference Manual,* Norwood, MA: Analog Devices, 1996.

13. P. H. Garrett, *Advanced Instrumentation and Computer I/O Design,* New York: IEEE Press, 1994.

14. D. H. Sheingold (Ed.), *Transducer Interfacing Handbook,* Norwood, MA: Analog Devices, 1980.

15. E. L. Zuch, *Data Acquisition and Conversion Handbook,* Mansfield, MA: Datel-Intersil, 1982.

16. H. W. Ott, *Noise Reduction Techniques in Electronic Systems,* New York: Wiley-Interscience, 1976.

17. G. Taguchi, *Introduction to Quality Engineering,* Tokyo: Asian Productivity Organization (JUSE), 1983.

18. Y. Akao, "Quality Function Deployment and CWQC in Japan," *Quality Progress,* October 1983, pp. 25–29.

19. V. Hunt. "Dual-Difference Redundant Structure In Fault-Tolerant Control," Aerospace Applications of Artificial Intelligence Conference, Dayton, OH, October 1989.

5

DATA CONVERSION DEVICES
AND ERRORS

5-0 INTRODUCTION

Data conversion devices provide the interfacing components between continuous-time signals representing the parameters of physical processes and their discrete-time digital equivalent. Recent emphasis on computer systems for automated manufacturing and the growing interest in using personal computers for data acquisition and control have increased the need for improved understanding of the design requirements of real-time computer I/O systems. However, before describing the theory and practice involved in these systems it is advantageous to understand the characterization and operation of the various devices from which these systems are fabricated. This chapter provides detailed information concerning A/D and D/A data conversion devices, and supporting components including analog multiplexers and sample-hold devices. The development of the individual error budgets representing these devices is also provided to continue the quantitative methodology of this text.

5-1 ANALOG MULTIPLEXERS

Field-effect transistors, both CMOS and JFET, are universally used as electronic multiplexer switches today, displacing earlier bipolar devices that had voltage offset problems. Junction FET switches have greater device electrical ruggedness and approximately the same switching speeds as CMOS devices. However, CMOS switches are dominant in multiplexer applications because of their unfailing turnoff, especially when the power is removed, unlike JFET devices, and their ability to multiplex signal levels up to the power supply voltages. Figure 5-1 shows a CMOS analog switch circuit where a stable ON resistance is achieved of about 100 Ω series resistance by the parallel p- and n-channel devices. Terminating a CMOS multiplexer with a high-input-impedance voltage follower eliminates any voltage divider errors possible as a consequence of the ON resistance. Figure 5-2 presents

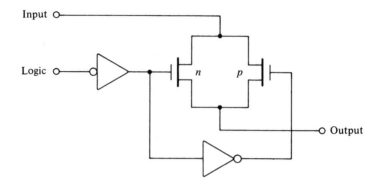

FIGURE 5-1. CMOS analog switch.

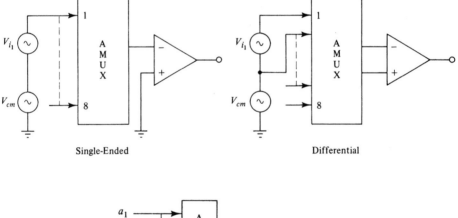

Single-Ended   Differential

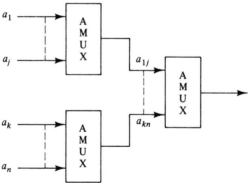

FIGURE 5-2. Multiplexer interconnections and tiered array.

TABLE 5-1. Multiplexer Switch Characteristics

Type	ON Resistance	OFF Isolation	Sample Rate
CMOS	100 Ω	70 dB	10 MHz
JFET	50 Ω	70 dB	1 MHz
Reed	0.1 Ω	90 dB	1 KHz

interconnection configurations for a multiplexer, and Table 5-1 lists multiplexer switch characteristics.

Errors associated with analog multiplexers are tabulated in Table 5-2, and are dominated by the average transfer error defined by equation (5-1). This error is essentially determined by the input voltage divider effect, and is minimized to a typical value of $\overline{0.01}$%FS when the AMUX is followed by an output buffer amplifier. The input amplifier associated with a sample-hold device often provides this high-impedance termination. Another error that can be significant is OFF-channel leakage current that creates an offset voltage across the input source resistance.

$$\text{Transfer error} = \frac{V_i - V_0}{V_i} \times 100\% \qquad (5\text{-}1)$$

5-2 SAMPLE-HOLDS

Sample-hold devices provide an analog signal memory function for use in sampled-data systems for temporary storage of changing signals for data conversion purposes. Sample-holds are available in several circuit variations, each suited to specific speed and accuracy requirements. Figure 5-3 shows a contemporary circuit that may be optimized either for speed or accuracy. The noninverting input amplifier provides a high-impedance buffer stage, and the overall unity feedback minimizes signal transfer error when the device is in the tracking mode. The clamping diodes ensure that the circuit remains stable during the hold mode when the switch is open. The inclusion of S/H devices in sampled-data systems must be carefully considered. The following examples represent the three essential applications for sample-holds. Table 5-3 lists representative sample-hold errors.

TABLE 5-2. Representative Multiplexer Errors

	REED	CMOS	
Transfer error	$\overline{0.01}$%	$\overline{0.01}$%	
Crosstalk error	0.001	0.001	
Leakage error		0.001	
Thermal offset	0.001		
$\varepsilon_{\text{AMUX}}$	0.01%FS	0.01%FS	$\overline{\Sigma\text{mean}} + 1\sigma$ RSS

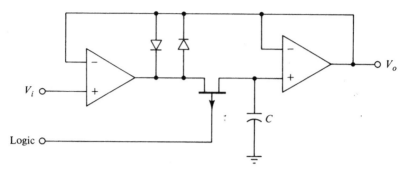

FIGURE 5-3. Closed-loop sample-hold.

Figure 5-4 diagrams a conventional multiplexed data conversion system cycle. The multiplexer and external circuit of Channel 1 are sampled by the S/H for a time sufficient for signal settling to within the amplitude error of interest. For sensor channels having RC time constants on the order of the S/H internal acquisition time, defined by equation (5-2), overlapping multiplexer channel selection and A/D conversion can speed system throughput significantly by means of an interposed sample-hold. A second application is described by Figure 5-5. Simultaneous data acquisition is required for many laboratory measurements in which multiple sensor channels must be acquired at precisely the same time. By matching S/H devices in bandwidth and aperture time, interchannel signal time skew can be minimized. The timing relationships are consequently preserved between signals, even though data conversion is performed sequentially.

$$\text{Acquisition time} = \frac{|V_0 - V_i|C}{I_o} + 9(R_o + R_{ON}) \, C \text{ seconds} \tag{5-2}$$

Voltage comparison A/D converters such as successive approximation devices require a constant signal value for accurate conversion. This function is normally provided by the application of a sample-hold preceding the AD converter, which constitutes the third application. An important issue is matching of S/H and A/D specifications to achieve the performance of interest. Sample-hold performance is

TABLE 5-3. Representative Sample-Hold Errors

Acquisition error	0.01%
Nonlinearity	0.004%
Gain	0.01%
Tempco	0.001%
$\varepsilon_{S/H}$	$\overline{\Sigma\text{mean}} + 1\sigma$ RSS 0.02%FS

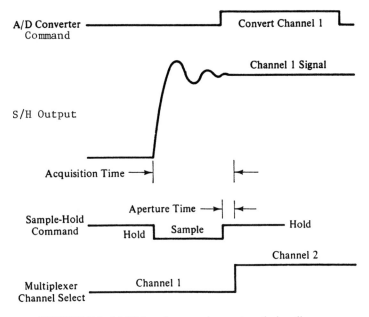

FIGURE 5-4. Multiplexed conversion system timing diagram.

principally determined by the input amplifier bandwidth and current output capability, which determines its ability to drive the hold capacitor C. A limiting parameter is the acquisition time of equation (5-2) and Figure 5-6, which when added to the conversion period T of an A/D converter determines the maximum throughput performance possible for a S/H and connected A/D. As a specific example, an Analog Devices 9100 device has an acquisition time of 14 ns for 0.01%FS (13-bit) settling, enabling data conversion rates to $(T + 14 \text{ ns})^{-1}$ Hz. In the sample mode, the charge

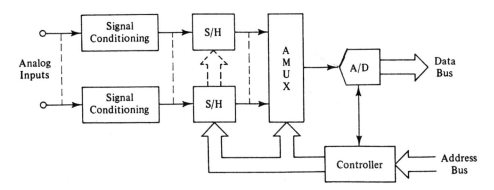

FIGURE 5-5. Simultaneous data acquisition.

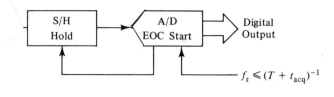

$$f_s \leqslant (T + t_{acq})^{-1}$$

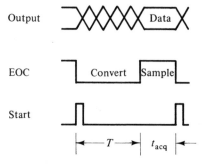

FIGURE 5-6. S/H-A/D Timing Relationships

on the hold capacitor is initially changed at the slew-limited output current capability I_o of the input amplifier. As the capacitor voltage enters the settling band coincident with the linear region of amplifier operation, final charging is exponential and corresponds to the summed time constants in equation (5-2), where R_o corresponds to amplifier output resistance and R_{ON} the switch resistance. The consequence of aperture time is to provide an average aperture error associated with the finite bound within which the amplitude of a sampled signal is acquired. Since this is a system error instead of a component error, its evaluation is deferred to Section 6-3.

5-3 DIGITAL-TO-ANALOG CONVERTERS

D/A converters, or DACs, provide reconstruction of discrete-time digital signals into continuous-time analog signals for computer interfacing output data recovery purposes such as actuators, displays, and signal synthesizers. D/A converters are considered prior to A/D converters because some AID circuits require DACs in their implementation. A D/A converter may be considered a digitally controlled potentiometer that provides an output voltage or current normalized to a full-scale reference value. A descriptive way of indicating the relationship between analog and digital conversion quantities is a graphical representation. Figure 5-7 describes a three-bit D/A converter transfer relationship having eight analog output levels ranging between zero and seven-eighths of full scale. Notice that a DAC full-scale digital input code produces an analog output equivalent to FS − 1 LSB. The basic structure of a conventional D/A converter includes a network of switched current

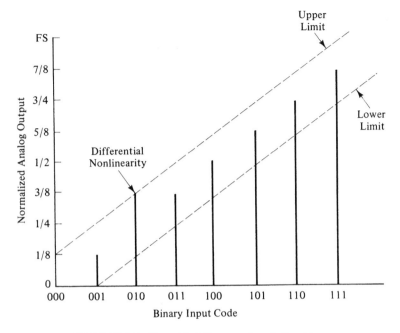

FIGURE 5-7. Three-bit D/A converter relationships.

sources having MSB to LSB values according to the resolution to be represented. Each switch closure adds a binary-weighted current increment to the output bus. These current contributions are then summed by a current-to-voltage converter amplifier in a manner appropriate to scale the output signal. Figure 5-8 illustrates such a structure for a three-bit DAC with unipolar straight binary coding corresponding to the representation of Figure 5-7.

In practice, the realization of the transfer characteristic of a D/A converter is nonideal. With reference to Figure 5-7, the zero output may be nonzero because of amplifier offset errors, the total output range from zero to FS − 1 LSB may have an overall increasing or decreasing departure from the true encoded values resulting from gain error, and differences in the height of the output bars may exhibit a curvature owing to converter nonlinearity. Gain and offset errors may be compensated for leaving the residual temperature-drift variations shown in Table 5-4 as the tempco of a representative 12-bit D/A converter. A voltage reference is necessary to establish a basis for the DAC absolute output voltage. The majority of voltage references utilize the bandgap principle, whereby the V_{be} of a silicon transistor has a negative tempco of −2 mV/°C that can be extrapolated to approximately 1.2 V at absolute zero (the bandgap voltage of silicon).

Converter nonlinearity is minimized through precision components, because it is essentially distributed throughout the converter network and cannot be eliminated by adjustment as with gain and offset errors. Differential nonlinearity and its varia-

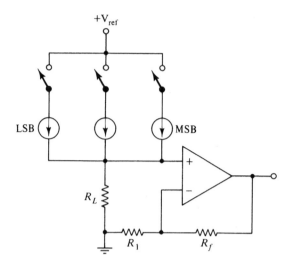

FIGURE 5-8. Straight binary three-bit DAC.

tion with temperature are prominent in data converters in that they describe the difference between the true and actual outputs for each of the 1 LSB code changes. A DAC with a 2 LSB output change for a 1 LSB input code change exhibits 1 LSB of differential nonlinearity as shown. Nonlinearities greater than 1 LSB make the converter output no longer single-valued, in which case it is said to be nonmonotonic and to have missing codes. Integral nonlinearity is an average error that generally does not exceed 1 LSB of the converter resolution as the sum of differential nonlinearities.

Table 5-5 presents frequently applied unipolar and bipolar codes expressed in terms of a 12-bit binary wordlength. These codes are applicable to both D/A and A/D converters. The choice of a code should be appropriate to the application and its sense understood (positive-true, negative-true). Positive-true coding defines a logic 1 as the positive logic level, and in negative-true coding the negative logic level is 1 with the other level 0. All codes utilized with data converters are based on the binary number system. Any base 10 number may be represented by equation (5-3), where the coefficient a_i assumes a value of 1 or 0 between the MSB (0.5) and LSB (2^{-n}). This coding scheme is convenient for data converters where the encoded

TABLE 5-4. Representative 12-Bit DAC Errors

Mean integral nonlinearity (1 LSB)	0.024%
Tempco (1 LSB)	0.024
Noise + distortion	0.001
$\varepsilon_{D/A}$	$\overline{\Sigma \text{mean}} + 1\sigma$ RSS 0.048%FS

TABLE 5-5. Data Converter Binary Codes

| | | Unipolar Codes—12-Bit Converters | | |
| | | Straight Binary and Complementary Binary | | |

Scale	+ 10 V FS	+ 5 V FS	Straight Binary	Complementary Binary
+ FS – 1 LSB	+ 9.9976	+ 4.9988	1111 1111 1111	0000 0000 0000
+ 7/8 FS	+ 8.7500	+ 4.3750	1110 0000 0000	0001 1111 1111
+ 3/4 FS	+ 7.5000	+ 3.7500	1100 0000 0000	0011 1111 1111
+ 5/8 FS	+ 6.2500	+ 3.1250	1010 0000 0000	0101 1111 1111
+ 1/2 FS	+ 5.0000	+ 2.5000	1000 0000 0000	0111 1111 1111
+ 3/8 FS	+ 3.7500	+ 1.8750	0110 0000 0000	1001 1111 1111
+ 1/4 FS	+ 2.5000	+ 1.2500	0100 0000 0000	1011 1111 1111
+ 1/8 FS	+ 1.2500	+ 0.6250	0010 0000 0000	1101 11111111
0 + 1 LSB	+ 0.0024	+ 0.0012	0000 0000 0001	1111 1111 1110
0	0.0000	0.0000	0000 0000 0000	1111 1111 1111

| | | BCD and Complementary BCD | | |

Scale	+ 10 V FS	+ 5 V FS	Binary Coded Decimal	Complementary BCD
+ FS – 1 LSB	+ 9.99	+ 4.95	1001 1001 1001	0110 0110 0110
+ 7/8 FS	+ 8.75	+ 4.37	1000 0111 0101	0111 1000 1010
+ 3/4 FS	+ 7.50	+ 3.75	0111 0101 0000	1000 1010 1111
+ 5/8 FS	+ 6.25	+ 3.12	0110 0010 0101	1001 1101 1010
+ 1/2 FS	+ 5.00	+ 2.50	0101 0000 0000	1010 1111 1111
+ 3/8 FS	+ 3.75	+ 1.87	0011 0111 0101	1100 1000 1010
+ 1/4 FS	+ 2.50	+ 1.25	0010 0101 0000	1101 1010 1111
+ 1/8 FS	+ 1.25	+ 0.62	0001 0010 0101	1110 1101 1010
0 + 1 LSB	+ 0.01	+ 0.00	0000 0000 0001	1111 1111 1110
0	0.00	0.00	0000 0000 0000	1111 1111 1111

| | | Bipolar Codes—12-Bit Converters | | | |

Scale	± 5 V FS	Offset Binary	Two's Complement	One's Complement	Sign-Magnitude Binary
+ FS – 1 LSB	+ 4.9976	1111 1111 1111	0111 1111 1111	0111 1111 1111	1111 1111 1111
+ 3/4 FS	+ 3.7500	1110 0000 0000	0110 0000 0000	0110 0000 0000	1110 0000 0000
+ 1/2 FS	+ 2.5000	1100 0000 0000	0100 0000 0000	0100 0000 0000	1100 0000 0000
+ 1/4 FS	+ 1.2500	1010 0000 0000	0010 0000 0000	0010 0000 0000	1010 0000 0000
0	0.0000	1000 0000 0000	0000 0000 0000	0000 0000 0000	1000 0000 0000
–1/4 FS	– 1.2500	0110 0000 0000	1110 0000 0000	1101 1111 1111	0010 0000 0000
–1/2 FS	– 2.5000	0100 0000 0000	1100 0000 0000	1011 1111 1111	0100 0000 0000
–3/4 FS	– 3.7500	0010 0000 0000	1010 0000 0000	1001 1111 1111	0110 0000 0000
– FS + 1 LSB	– 4.9976	0000 0000 0001	1000 0000 0001	1000 0000 0000	0111 1111 1111
– FS	– 5.0000	0000 0000 0000	1000 0000 0000		

value is interpreted in terms of a fraction of full scale for n-bit word lengths. Straight-binary, positive-true unipolar coding is most commonly encountered. Complementary binary positive-true coding is identical to straight binary negative-true coding. Sign-magnitude bipolar coding is often used for outputs that are frequently in the vicinity of zero. Offset binary is readily converted to the more computer-compatible two's complement code by complementing the MSB.

$$N = \sum_{i=0}^{n} a_i 2^{-i} \tag{5-3}$$

As the input code to a DAC is increased or decreased, it passes through major and minor transitions. A major transition is at half-scale when the MSB is switched and all other switches change state. If some switched current sources lag others, then significant transient spikes known as glitches are generated. Glitch energy is of concern in fast-switching DACs driven by high-speed logic with time skew between transitions. However, high-speed DACs also frequently employ an output S/H circuit to deglitch major transitions by remaining in the hold mode during these intervals. Internally generated noise is usually not significant in D/A converters except at extreme resolutions, such as the 20-bit Analog Devices DAC 1862, whose LSB is equal to 10 μ V with 10 V_{FS} scaling.

The advent of monolithic D/A converters has resulted in almost universal acceptance of the $R - 2R$ network DAC because of the relative ease of achieving precise resistance ratios with monolithic technology. This is in contrast to the low yields experienced with achieving precise absolute resistance values required by weighted-resistor networks. Equations (5-4) and (5-5) define the quantities of each converter. For the $R - 2R$ network, an effective resistance of 3 R is seen by V_{ref} for each branch connection with equal left–right current division (see Figure 5-9).

$$V_0 = \frac{R_f}{R} \cdot V_{ref} \cdot \sum_{i=0}^{n} 2^{-i} \qquad \text{Weighted} \tag{5-4}$$

$$V_0 = \frac{R_f}{2R} \cdot \frac{V_{ref}}{3} \cdot \sum_{i=0}^{n} 2^{-i} \qquad R - 2R \tag{5-5}$$

A D/A converter that accepts a variable reference can be configured as a multiplying DAC that is useful for many applications requiring a digitally controlled scale factor. Both linear and logarithmic scale factors are available for applications such as, respectively, digital excitation in test systems and a dB step attenuator in communications systems. The simplest devices operate in one quadrant with a unipolar reference signal and digital code. Two-quadrant multiplying DACs utilize either bipolar reference signals or bipolar digital codes. Four-quadrant multiplication involves both a bipolar reference signal and bipolar digital code. Table 5-6 describes a two-quadrant, 12-bit linear multiplying D/A converter. The variable transconductance property made possible by multiplication is useful for many signal conditioning applications, including programmable gain.

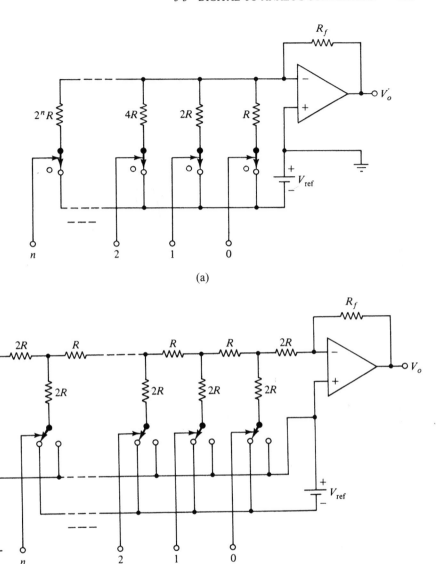

FIGURE 5-9. (a) Weighted resistor D/A converter. (b) $R - 2R$ resistor D/A converter.

As system peripheral complexity has expanded to require more of a host computer's resources, peripheral interface devices have been provided with transparent
processing capabilities to more efficiently distribute these tasks. In fact, some devices such as video and graphics processors are more complicated than the host
computer they support. Universal peripheral bus master devices have evolved that
offer a flexible combination of memory-mapped, interrupt-driven, and DMA data

TABLE 5-6. Two-Quadrant Multiplying 12-Bit DAC

Straight Binary Input	Analog Output
1111 1111 1111	$\pm V_i\left(\dfrac{4095}{4096}\right)$
1000 0000 0001	$\pm V_i\left(\dfrac{2048}{4096}\right)$
0000 0000 0001	$\pm V_i\left(\dfrac{1}{4096}\right)$
0000 0000 0000	0 V

transfer capabilities with FIFO buffer memory for accommodation of multiple buses and differing speeds. The example D/A peripheral interface of Figure 5-10 employs a program-initiated output whose status is polled by the host for a Ready enable. Data may then be transferred to the D port with $\overline{\text{IOW}}$ low and $\overline{\text{CE}}$ high.

5-4 ANALOG-TO-DIGITAL CONVERTERS

The conversion of continuous-time analog signals to discrete-time digital signals is fundamental to obtaining a representative set of numbers that can be used by a digi-

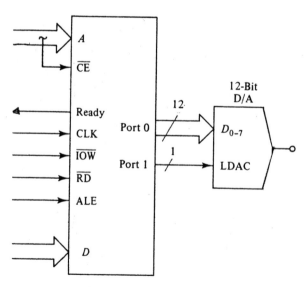

FIGURE 5-10. D/A peripheral interface.

tal computer. The three functions of sampling, quantizing, and encoding are involved in this process and implemented by all A/D converters, as illustrated by Figure 5-11. The detailed system considerations associated with these functions and their relationship to computer interface design are developed in Chapter 6. We are concerned here with A/D converter devices and their functional operations, as we were with the previously described data conversion devices. In practice, one conversion is performed each period T, the inverse of sample rate f_s, whereby a numerical value derived from the converter quantizing levels is translated to an appropriate output code. The graph of Figure 5-12 describes A/D converter input–output relationships and quantization error for prevailing uniform quantization, where each of the levels q is of spacing 2^{-n} (1 LSB) for a converter having an n-bit binary output wordlength. Note that the maximum output code does not correspond to a full-scale input value, but instead to $(1 - 2^{-n}) \cdot$ FS because there exist only $(2^{-n} - 1)$ coding points, as shown in Figure 5-12.

Quantization of a sampled analog waveform involves the assignment of a finite number of amplitude levels corresponding to discrete values of input signal V_s between 0 and V_{FS}. The uniformly spaced quantization intervals 2^{-n} represent the resolution limit for an n-bit converter, which may also be expressed as the quantizing interval q equal to $V_{FS}/(2^{-n} - 1)V$. Figure 5-13 illustrates the prevailing uniform quantizing algorithm whereby an input signal that falls within the V_jth-level range of $\pm q/2$ is encoded at the V_jth level with a quantization error of ε volts. This error may range up to $\pm q/2$, and is an irreducible noise added to a converter output signal. The conventional assumption concerning the probability density function of this noise is that it is uniformly distributed along the interval $\pm q/2$, and is represented as the A/D converter quantizing uncertainty error of value 1/2 LSB proportional to converter wordlength.

The equivalent rms error of quantization (E_{qe}) produced by this noise is described by equation (5-6). The rms sinusoidal signal-to-noise ratio (SNR) of equation (5-7) then defines the output signal quality achievable, expressed in power dB, for an A/D converter of n bits with a noise-free input signal. These relationships are tabulated in Table 5-7. Equation (5-8) defines the dynamic range of a data converter of n bits in voltage dB. Converter dynamic range is useful for matching A/D converter wordlength in bits to a required analog input signal span to be represented digitally. For example, a 10 mV-to-10 V span (60 voltage dB) would require a min-

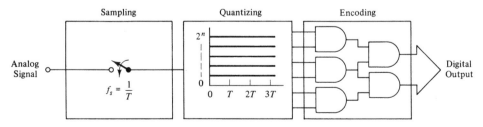

FIGURE 5-11. A/D converter functions.

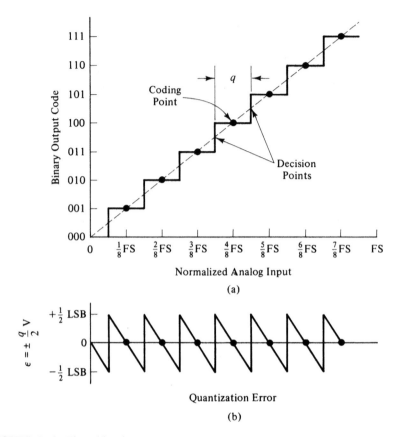

FIGURE 5-12. Three-bit A/D converter relationships: (a) quantization intervals; (b) quantization error.

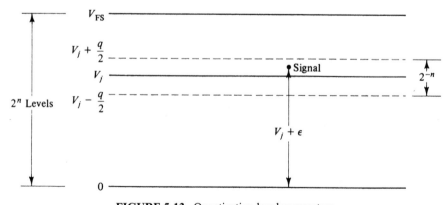

FIGURE 5-13. Quantization level parameters.

TABLE 5-7. Decimal Equivalents of 2^n and 2^{-n}

Bits, n	Levels, 2^n	LSB Weight, 2^{-n}	Quantization SNR, dB
1	2	0.5	8
2	4	0.25	14
3	8	0.125	20
4	16	0.0625	26
5	32	0.03125	32
6	64	0.015625	38
7	128	0.0078125	44
8	256	0.00390625	50
9	512	0.001953125	56
10	1,024	0.0009765625	62
11	2,048	0.00048828125	68
12	4,096	0.000244140625	74
13	8,192	0.0001220703125	80
14	16,384	0.00006103515625	86
15	32,768	0.000030517578125	92
16	65,536	0.0000152587890625	98
17	131,072	0.00000762939453125	104
18	262,144	0.000003814697265625	110
19	524,288	0.0000019073486328125	116
20	1,048,576	0.00000095367431640625	122

imum converter wordlength n of 10 bits. It will be shown in Section 6-3 that additional considerations are involved in the conversion of an input signal to an n-bit accuracy other than the choice of A/D converter wordlength, where the dynamic range of a digitized signal may be represented to n bits without achieving n-bit data accuracy. However, the choice of a long wordlength A/D converter will beneficially minimize both quantization noise and A/D device error and provide increased converter linearity.

$$\text{Quantization error } E_{qe} = \left(\frac{1}{q} \int_{-q/2}^{q/2} \varepsilon^2 \cdot d\varepsilon \right)^{1/2} \tag{5-6}$$

$$= \frac{q}{2\sqrt{3}} \text{ rms volts}$$

$$\text{Quantization SNR} = 10 \log \left(\frac{V_{FS}/2\sqrt{2}}{E_{qe}} \right)^2 \tag{5-7}$$

$$= 10 \log \left(\frac{2^n \cdot q/2\sqrt{2}}{q/2\sqrt{3}} \right)^2$$

$$= 6.02n + 1.76 \text{ power dB}$$

$$\text{Dynamic range} = 20 \log (2^n) \tag{5-8}$$

$$= 6.02n \text{ voltage dB}$$

The input comparator is critical to the conversion speed and accuracy of an A/D converter as shown in Figure 5-14. Generally, it must possess sufficient gain and bandwidth to achieve switching and settling to the amplitude error of interest ultimately determined by noise sources present, such as described in Section 4-1.

Described now are seven prevalent A/D conversion methods and their application considerations. Architectures presented include integrating dual-slope, sampling successive-approximation, digital angle converters, charge-balancing and its evolution to oversampling sigma–delta converters, simultaneous or flash, and pipelined subranging. The performance of these conversion methods all benefit from circuit advances and monolithic technologies in their accuracy, stability, and reliability that permit expression in terms of simplified static, dynamic, and temperature parameter error budgets, as illustrated by Table 5-8.

Quantizing uncertainty constitutes converter dynamic amplitude error, illustrated by Figure 5-12(b). Mean integral nonlinearity describes the maximum deviation of the static-transfer characteristic between initial and final code transitions in Figure 5-12(a). Circuit offset, gain, and linearity temperature coefficients are combined into a single percent of full-scale tempco expression. Converter signal-to-noise plus distortion expresses the quality of spurious and linearity dynamic performance. This latter error is influenced by data converter -3 dB frequency response, which generally must equal or exceed its conversion rate f_s to avoid amplitude and phase errors, considering the presence of input signal BW values up to the $f_s/2$ Nyquist frequency and prudent response provisions. It is notable from Table 5-8 that the sum of the mean and RSS of other converter errors provides a digital accuracy whose effective number of bits is typically less than the specified converter wordlength.

Integrating converters provide noise rejection for the input signal at an attenuation rate of -20 dB/decade of frequency, as described in Figure 5-15, with sinc nulls at multiples of the integration period T by equation (5-9). The ability of an integra-

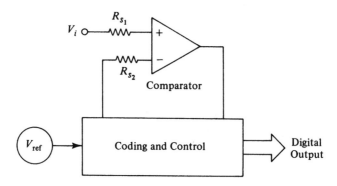

FIGURE 5-14. Comparator-oriented A/D converter diagram.

TABLE 5-8. Representative 12-Bit ADC Errors

Mean integral nonlinearity (1 LSB)	$\overline{0.024\%}$
Quantizing uncertainty ($\frac{1}{2}$ LSB)	0.012
Tempco (1 LSB)	0.024
Noise + distortion	0.001
$\varepsilon_{A/D}$	$\overline{\Sigma\text{mean}} + 1\sigma$ RSS 0.050%FS

tor to provide this response is evident from its frequency response $H(\omega)$, obtained by the integration of its impulse response $h(t)$ in equation (5-9). Note that this noise improvement requires integration of the signal plus noise during the conversion period, and consequently is not furnished when a sample-hold device precedes the converter. A conversion period of $16\frac{2}{3}$ ms will provide a useful null to 60 Hz interference, for example.

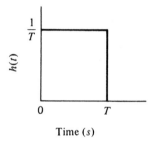

Time (s)

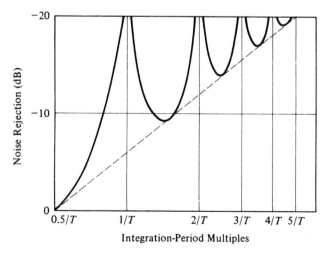

FIGURE 5-15. Integrating converter noise rejection.

$$H(\omega) = \int_0^T h(t) \cdot e^{-j\omega t} \cdot dt \qquad (5\text{-}9)$$

$$= e^{-j\omega T/2} \cdot \frac{\sin \omega T/2}{\omega T/2}$$

Integrating dual-slope converters perform A/D conversion by the indirect method of converting an input signal to a representative pulse sequence that is totaled by a counter. Features of this conversion technique include self-calibration to component temperature drift, use of inexpensive components in its mechanization, and multiphasic integrations yielding improved resolution of the zero endpoint shown in Figure 5-16. Operation occurs in three steps. First, the autozero phase stores converter analog offsets on the integrator with the input grounded. Second, an input signal is integrated for a fixed time T_1. Finally, the input is connected to a reference of opposite polarity and integration proceeds to zero during a variable time T_2 within which clock pulses are totaled in proportion to the input signal amplitude. These operations are described by equations (5-10) and (5-11). Integrating

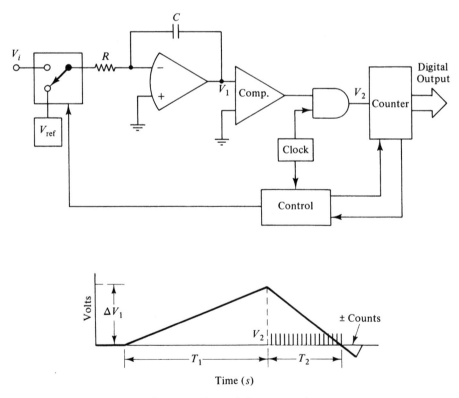

FIGURE 5-16. Dual-slope conversion.

converters are early devices whose merits are best applied to narrow bandwidth signals such as encountered with hand-held multimeters. Wordlengths to 16 bits are available, but conversion is limited to 1 KSPS.

$$\Delta V_1 = \frac{1}{RC} \cdot V_i \cdot T_{1\text{constant}} \tag{5-10}$$

$$= \frac{1}{RC} \cdot V_{\text{ref}} \cdot T_{2\text{variable}}$$

$$T_2 = \frac{V_i \cdot T_1}{V_{\text{ref}}} \tag{5-11}$$

The successive approximation technique is the most widely applied A/D converter type for computer interfacing, primarily because its constant conversion period T is independent of input signal amplitude. However, it requires a preceding S/H to satisfy its requirement for a constant input signal. This feedback converter operates by comparing the output of an internal D/A converter with the input signal at a comparator, where each bit of the wordlength is sequentially tested during n equal time subperiods in the development of an output code representative of input signal amplitude. Converter linearity is determined by the performance of its internal D/A. Figure 5-17 describes the operation of a sampling successive approximation converter. The conversion period and S/H acquisition time combined determine the maximum conversion rate as described in Figure 5-6. Successive approximation converters are well suited for converting arbitrary signals, including those that are nonperiodic, in multiplexed systems. Wordlengths of 16 bits are available at conversion rates to 500 KSPS.

A common method for representing angles in digital form is in natural binary weighting, where the most significant bit (MSB) represents 180 degrees and the MSB- 1 represents 90 degrees, as tabulated in Table 5-9. Digital synchro conversion shown in Figure 5-18 employs a Scott-T transformer connection and ac reference to develop the signals defined by equations (5-12) and (5-13). Sine ϕ and cosine ϕ quadrature multiplications are achieved by multiplying-D/A converters whose difference is expressed by equation (5-14). A phase-detected dc error signal, described by equation (5-15), then pulses an up/down counter to achieve a digital output corresponding to the sinchro angle θ. Related devices include digital vector generators that generate quadrature circular functions as analog outputs from digital angular inputs.

$$V_A = \sin(377t) \cdot \sin \theta \tag{5-12}$$

$$V_B = \sin(377t) \cdot \cos \theta \tag{5-13}$$

$$V_E = \sin(377t) \cdot \sin(\theta - \phi) \tag{5-14}$$

$$V_D = \sin(\theta - \phi) \tag{5-15}$$

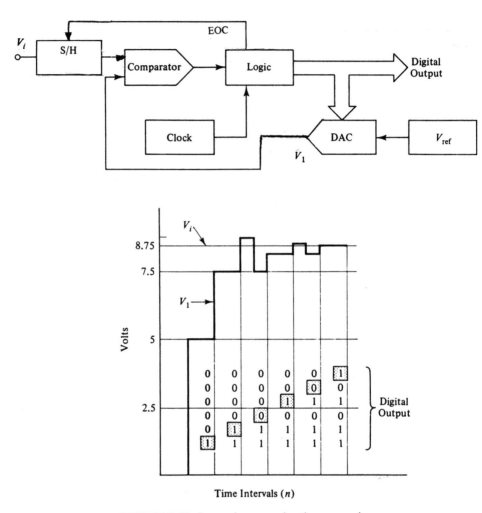

FIGURE 5-17. Successive approximation conversion.

Charge-balancing A/D converters utilize a voltage-to-frequency circuit to convert an input signal to a current I_i from which is subtracted a reference current I_{ref}. This difference current is then integrated for successive intervals, with polarity reversals determined in one direction by a threshold comparator and in the other by clock count. The conversion period for this converter is constant, but the number of count intervals per conversion vary in direct proportion to input signal amplitude, as illustrated in Figure 5-19. Although the charge-balancing converter is similar in performance to the dual-slope converter, their applications diverge; the former is compatible with and integrated in microcontroller devices.

Sigma–delta conversion employs a version of the charge-balancing converter as its first stage to perform one-bit quantization at an oversampled conversion rate f_s

TABLE 5-9. Binary Angle Representation

Bit	Degrees
1	180
2	90
3	45
4	22.5
5	11.25
6	5.625
7	2.812
8	1.406
9	0.703
10	0.351
11	0.176
12	0.088

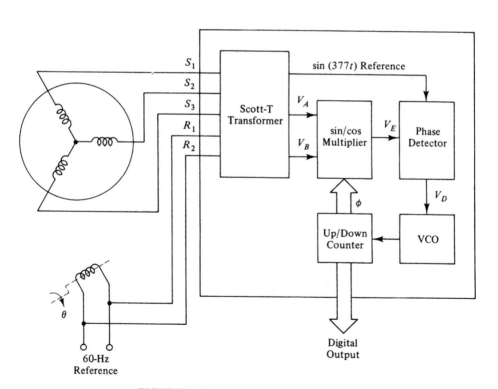

FIGURE 5-18. Synchro-to-digital conversion.

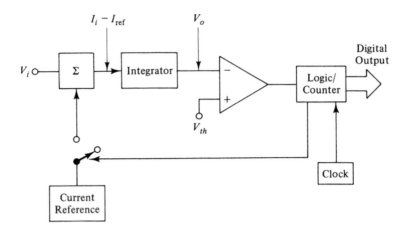

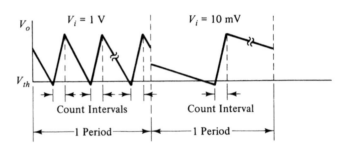

FIGURE 5-19. Charge-balancing conversion.

whose "ones" density corresponds to analog input signal amplitude. The high quantizing noise resulting from one-bit conversion is effectively spread over a wide bandwidth from the oversampling operation, which is amenable to efficient digital filtering since it is in the digital domain. The resulting spectrum is then resampled to an equivalent Nyquist-sampled signal BW of n-bit resolution shown in Figure 5-20. Sigma–delta converters are prevalent in medium-bandwidth, high-resolution periodic signal applications from measurement instruments to telecommunications and consumer electronics. Word lengths to 20 bits for 100 kHz signal BW are available. Because of signal latency associated with oversampling and decimation operations, however, sigma–delta converters are not compatible with multiplexed applications that encounter nonperiodic or transient signals.

Simultaneous or flash converters are represented by the diagram of Figure 5-21 and require 2^n-1 comparators biased 1 LSB apart to encode an analog input signal to n-bit resolution. All quantization levels are simultaneously compared in a single clock cycle that produces a comparator "thermometer" code with a 1/0 boundary proportional to input signal amplitude. Comparator coding logic then provides a fi-

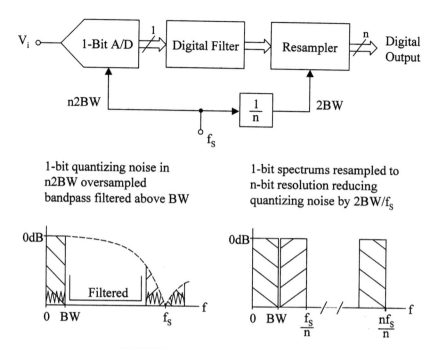

FIGURE 5-20. Sigma–delta conversion.

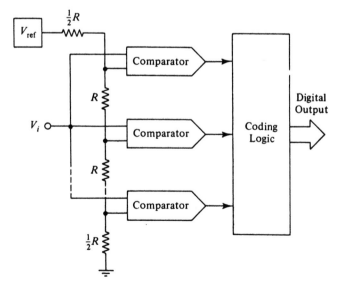

FIGURE 5-21. Flash converter.

nal digital output word. This architecture offers the fastest conversion rate achievable in a single clock cycle, but resolution is practically limited by the increasing number of comparators required for extending output wordlength. Wordlengths to 10 bits with 1023 comparators are available, however, at real-time rates to 100 MSPS. The flash converter beneficially can accommodate dynamic nonperiodic signals, like the slower successive approximation converter, without an input S/H device. Applications include radar processors, electro-optical systems, and professional video.

Wideband and widerange conversion is the province of pipelined subranging converters, which offer higher resolution than flash converters with nearly the same conversion rates. Wordlengths of 12 bits are common at conversion rates to 80 MSPS for applications ranging from digital spectrum analyzers to medical imaging. This architecture overcomes the comparator limitation of flash converters by dividing the conversion task into cascaded stages. A typical two-stage subranging converter is shown in Figure 5-22 with two six-bit flash A/D converters requiring only 126 comparators to provide a 12-bit wordlength, where the differential subrange is converted to LSB values by the second A/D. Flash converters of m bits in p stages offer a resolution of $p \times m$ bits with $p \times (2^m-1)$ comparators. The throughput latency of p cycles of the pipeline impedes the conversion of nonperiodic signals, however.

Interrupt-initiated interfacing provides the flexibility required to accommodate asynchronous inputs from A/D converters. Upon the request of a peripheral controller, a processor interrupt is generated that initiates a service routine containing the device handler. This structure is illustrated by Figure 5-23 and offers enhanced throughput in comparison with program-initiated interfacing, previously described for D/A converters, by reconciling speed differences between processor buses and their peripheral controllers with FIFO buffers and DMA data transfers. Vectored multilevel-priority interrupts are also readily accommodated by this structure.

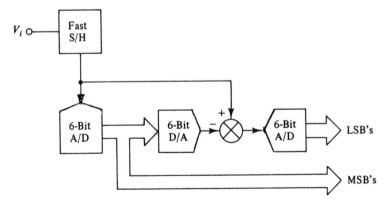

FIGURE 5-22. Wideband 12-bit subranging A/D converter.

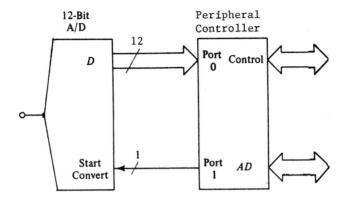

FIGURE 5-23. Interrupt-initiated A/D interface.

BIBLIOGRAPHY

1. *Analog-Digital Conversion Handbook,* Norwood, MA: Analog Devices, 1972.

2. E. R. Hnatek, *A User's Handbook of D/A and A/D Converters,* New York: Wiley, 1976.

3. B. M. Gordon., *The Analogic Data-Conversion Systems Digest,* Wakefield, MA: Analogic, 1977.

4. E. Zuch, *Data Acquisition and Conversion Handbook,* Mansfield, MA: Datel-Intersil, 1977.

5. D. Stantucci, "Data Acquisition Can Falter Unless Components are Well Understood," *Electronics,* November 27, 1975.

6. B. M. Gordon, "The ABC's of A/D and D/A Converter Specifications," *Electronic Design News,* August 1972.

7. M. Lindheimer, "Guidelines for Digital-to-Analog Converter Applications," *Electronic Equipment Engineering,* September 1970.

8. E. Zuch, "Consider V/F Converters," *Electronic Design,* November 22, 1976.

9. R. Allen, "A/D and D/A Converters: Bridging the Analog World to the Computer," *Electronic Design News,* February 5, 1973.

10. B. A. Artwick, *Microcomputer Interfacing,* Englewood Cliffs, NJ: Prentice Hall, 1980.

11. D. F. Hoeschele, *Analog-to-Digital, Digital-to-Analog Conversion Techniques,* New York: Wiley, 1968.

12. G. E. Tobey, "Ease Multiplexing and A/D Conversion," *Electronic Design,* April 12, 1973.

13. Synchro and Resolver Conversion, Norwood, MA: Analog Devices, 1980.

6

SAMPLING AND RECONSTRUCTION WITH INTERSAMPLE ERROR

6-0 INTRODUCTION

A fundamental requirement of sampled-data systems is the sampling of continuous-time signals to obtain a representative set of numbers that can be used by a digital computer. The primary goal of this chapter is to provide an understanding of this process. The first section explores theoretical aspects of sampling and the formal considerations of signal recovery, including ideal Wiener filtering in signal interpolation. Aliasing of signal and noise are considered next in a detailed development involving a heterodyne basis of evaluation. This development coordinates signal bandwidth, sample rate, and band-limiting prior to sampling to achieve minimum aliasing error under conditions of significant aliasable content. The third section addresses intersample error in sampled systems, and provides a sample-rate-to-signal-bandwidth ratio (f_s/BW) expressing the step-interpolator representation of sampled data in terms of equivalent binary accuracy. The final section derives a mean-squared error criterion for evaluating the performance of practical signal recovery techniques. This provides an interpolated output signal accuracy in terms of the corresponding minimum required sample rate and suggests a data conversion system design procedure that is based on considering system output performance requirements first.

6-1 SAMPLED DATA THEORY

Observation of typical sensor signals generally reveals band-limited continuous functions with a diminished amplitude outside of a specific frequency band, except for interference or noise, which may extend over a wide bandwidth. This is attributable to the natural roll-off or inertia associated with actual processes or systems providing the sensor excitation. Sampled-data systems provide discrete signals of finite accuracy from continuous signals of true accuracy. Of interest is how much information is lost by the sampling operation and to what accuracy an original continuous signal can be reconstructed from its sampled values. The consideration of periodic sampling offers a mathematical solution to this problem for band-limited sampled signals of bandwidth BW. Signal discretization is illustrated for the two classifications of nonreturn-to-zero (NRZ) sampling and return-to-zero (RZ) sampling in Figure 6-1. This figure represents the two sampling classifications in both the time and frequency domains, where τ is the sampling function width and T the sampling period (the latter the inverse of sample rate f_s). The determination of specific sample rates that provide sampled-data accuracies of interest is a central theme of this chapter.

The provisions of periodic sampling are based on Fourier analysis and include the existence of a minimum sample rate for which theoretically exact signal reconstruction is possible from the sampled sequence. This is significant in that signal sampling and recovery are considered simultaneously, correctly implying that the design of data conversion and recovery systems should also be considered jointly. The interpolation formula of equation (6-1) analytically describes the approximation $\hat{x}(t)$ of a continuous-time signal $x(t)$ with a finite number of samples from the sequence $x(nT)$. $\hat{x}(t)$ is obtained from the inverse Fourier transform of the input sequence, which is derived from $x(t) \cdot \hat{p}(t)$ as convolved with the ideal interpolation function $H(f)$ of Figure 6-2. This results in the sinc amplitude response in the time domain owing to the rectangular characteristic of $H(f)$. Due to the orthogonal behavior of equation (6-1) only one nonzero term is provided at each sampling instant. Contributions of samples other than ones in the immediate neighborhood of a specific sample diminish rapidly because the amplitude response of $H(f)$ tends to decrease inversely with the value of n. Consequently, the interpolation formula provides a useful relationship for describing recovered band-limited sampled-data signals, with T chosen sufficiently small to prevent signal aliasing. Aliasing is discussed in detail in the following section. Figure 6-3 shows the behavior of this interpolation formula including its output approximation $\hat{x}(t)$.

$$\hat{x}(t) = F^{-1}\{f[x(nT)] \cdot H(f)\} \tag{6-1}$$

$$= \sum_{n=-x}^{x} \left\{ T \int_{-BW}^{BW} x(nT)e^{-j2\pi fnT} \right\} \cdot e^{j2\pi ft} \cdot df$$

$$= T \sum_{n=-x}^{x} x(nT) \frac{e^{j2\pi BW(t-nT)} - e^{-j2\pi BW(t-nT)}}{j2\pi(t-nT)}$$

$$= 2TBW \sum_{n=-x}^{x} x(nT) \frac{\sin 2\pi BW(t-nT)}{2\pi BW(t-nT)}$$

A formal description of this process was provided both by Wiener [13] and Kolmogoroff [15]. It is important to note that the ideal interpolation function $H(f)$ uti-

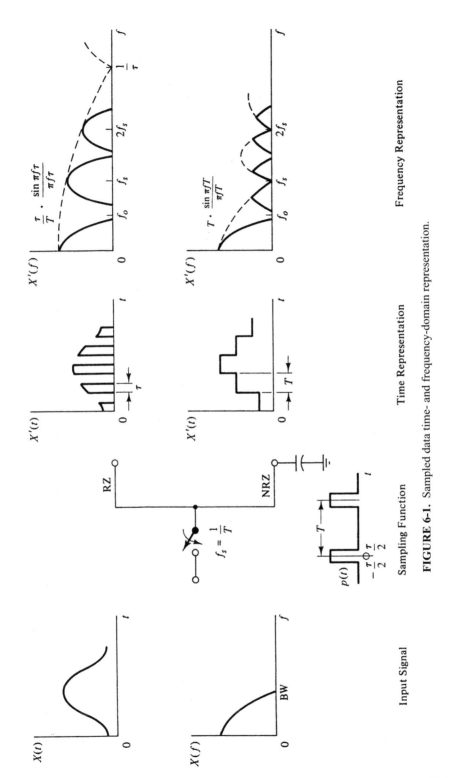

FIGURE 6-1. Sampled data time- and frequency-domain representation.

123

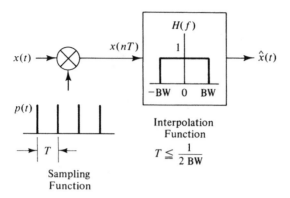

FIGURE 6-2. Ideal sampling and recovery.

lizes both phase and amplitude information in reconstructing the recovered signal $\hat{x}(t)$, and is therefore more efficient than conventional linear filters. However, this ideal interpolation function cannot be physically realized because its impulse response $H(f)$ is noncausal, requiring an output that anticipates its input. As a result, practical interpolators for signal recovery utilize amplitude information that can be made efficient, although not optimum, by achieving appropriate weighting of the reconstructed signal. These principles are observed Section 6-4.

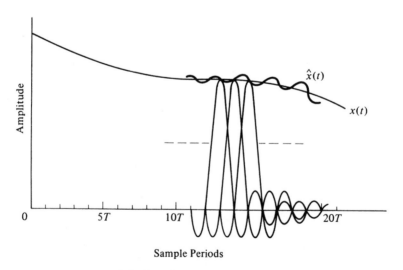

FIGURE 6-3. Signal interpolation.

A significant consideration imposed upon the sampling operation results from the finite width τ of practical sampling functions, denoted by $p(t)$ in Figure 6-1. Since the spectrum of a sampled signal consists of its original baseband spectrum $X(f)$ plus a number of images of this signal, these image signals are shifted in frequency by an amount equal to the sampling frequency f_s and its harmonics mf_s as a consequence of the periodicity of $p(t)$. The width of τ determines the amplitude of these signal images, as attenuated by the sinc functions described by the dashed lines of $X'(f)$ in Figure 6-1, for both RZ and NRZ sampling. Of particular interest is the attenuation impressed upon the baseband spectrum of $X'(f)$ corresponding to the amplitude and phase of the original signal $X(f)$. A useful criterion is to consider the average baseband amplitude error between dc and the signal BW expressed as a percentage of the full-scale departure from unity gain. Also, digital processor bandwidth must be sufficient to support these image spectra until their amplitudes are attenuated by the sinc function to preserve signal fidelity. The mean sinc amplitude error is expressed for RZ and NRZ sampling by equations (6-2) and (6-3). The sampled-data bandwidth requirement for NRZ sampling is generally more efficient in system bandwidth utilization than the $1/\tau$ null provided by RZ sampling. The minimization of mean sinc amplitude error may also influence the choice of f_s. The folding frequency f_o in Figure 6-1 is an identity equal to $f_s/2$, and the specific NRZ sinc attenuation at f_o is always 0.636, or -3.93 dB.

$$\overline{\varepsilon_{\text{RZ sinc \%FS}}} = \frac{1}{2}\left(1 - \frac{\tau}{T} \cdot \frac{\sin \pi BW\tau}{\pi BW\tau}\right) \cdot 100\% \tag{6-2}$$

$$\overline{\varepsilon_{\text{RZ sinc \%FS}}} = \frac{1}{2}\left(1 - \frac{\sin \pi BWT}{\pi BWT}\right) \cdot 100\% \tag{6-3}$$

RZ sampling is primarily used for multiplexing multichannel signals into a single channel, such as encountered in telemetry systems. Figure 6-1 provides that the dc component of RZ sampling has an amplitude of τ/T, its average value or sampling duty cycle, which may be scaled as required by the system gain. NRZ sampling is inherent in the operation of all data-conversion components encountered in computer input–output systems, and reveals a dc component proportional to the sampling period T. In practice, this constant is normalized to unity by the $1/T$ impulse response associated with the transfer functions of actual data-conversion components.

Note that the sinc function and its attenuation with frequency in a sampled-data system is essentially determined by the duration of the sampled-signal representation $X'(t)$ at any point of observation, as illustrated in Figure 6-1. For example, an A/D converter with a conversion period T double the value employed for a following connected D/A converter will exhibit an NRZ sinc function having twice the attenuation rate versus frequency as that of the D/A, which is attributable to the transformation of the sampled-signal duration. D/A oversampling accordingly offers reduced output sinc error, illustrated by Figure 6-15.

6-2   ALIASING OF SIGNAL AND NOISE

The effect of undersampling a continuous signal is illustrated in both the time and frequency domains in Figure 6-4. This demonstrates that the mapping of a signal to its sampled-data representation does not have an identical reverse mapping if it is reconstructed as a continuous signal when it is undersampled. Such signals appear as lower-frequency aliases of the original signal, and are defined by equation (6-4) when $f_s < 2\ BW$. As the sample rate f_s is reduced. samples move further apart in the time domain, and signal images closer together in the frequency domain. When image spectrums overlap, as illustrated in Figure 6-4b, signal aliasing occurs. The consequence of this result is the generation of intermodulation distortion that cannot be removed by later signal processing operations. Of interest is aliasing at f_o between the baseband spectrum, representing the amplitude and phase of the original signal, and the first image spectrum. The folding frequency f_o is the highest frequency at which sampled-data signals may exist without being undersampled. Accordingly, f_s must be chosen greater than twice the signal BW to ensure the absence of signal aliasing, which usually is readily achieved in practice.

$$f_{alias} = [f_s - BW] \qquad f_s < 2\ BW$$
$$= \text{nonexistent} \qquad f_s \geq 2\ BW \qquad (6\text{-}4)$$

Of greater general concern and complexity is noise aliasing in sampled-data systems. This involves either out-of-band signal components, such as coherent inter-

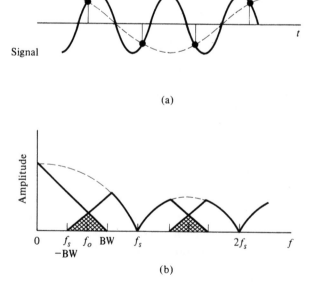

(a)

(b)

FIGURE 6-4. Time (a) and frequency (b) representation of undersampled signal aliasing.

ference or random noise spectra, present above f_o and therefore undersampled. One or more of these sources are frequently present in most sampled-data systems. Consequently, the design of these systems should provide for the analysis of noise aliasing and the coordination of system parameters to achieve the aliasing attenuation of interest. Understanding of baseband aliasing is aided with reference to Figures 6-5 and 6-6. The noise aliasing source bands shown are heterodyned within the baseband signal between dc and f_o, derived by equation (6-5) as $mf_s - BW \leq f_{\text{noise}} < mf_s + BW$, as a consequence of the sampling function spectra, which arise at multiples of f_s. The resulting combination of signal and aliasing components generate intermodulation distortion proportional to the baseband alias amplitude error derived by equations (6-6) through (6-10).

$$mf_s - BW \leq f_{\text{noise}} < mf_s + BW \qquad \text{alias source frequencies} \qquad (6\text{-}5)$$

$$f_{\text{coherent alias}} = |mf_s - f_{\text{coh}}| \qquad \text{at baseband} \qquad (6\text{-}6)$$

$$= 24 \text{ Hz} - 23 \text{ Hz}$$

$$= 1 \text{ Hz } (m = 1)$$

$$\varepsilon_{\text{coherent alias}} = V_{\text{coh\%FS}} \cdot \text{filter attn} \cdot \text{sinc} \qquad (6\text{-}7)$$

$$= 50\%\text{FS} \cdot \frac{1}{\sqrt{1 + \left(\dfrac{f_{\text{coh}}}{f_c}\right)^{2n}}} \cdot \text{sinc}\left(\frac{|mf_s - f_{\text{coh}}|}{f_s}\right)$$

$$= 50\%\text{FS} \cdot \frac{1}{\sqrt{1 + \left(\dfrac{23}{3}\right)^{6}}} \cdot \text{sinc}\left(\frac{|24 - 23|}{24}\right)$$

$$= 50\%\text{FS} \cdot (0.0024) \cdot (0.998)$$

$$= 0.12\%\text{FS with presampling filter}$$

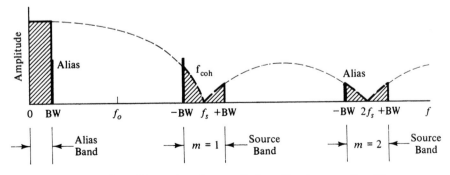

FIGURE 6-5. Coherent interference aliasing without presampling filter.

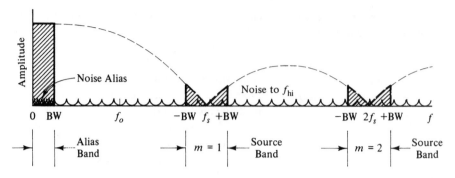

FIGURE 6-6. Random interference aliasing without presampling filter.

$$N_{\text{alias}} = \overset{\text{\# source bands}}{\underset{0}{\sum}} (V_{\text{noise}} \text{ rms})^2 \cdot (\text{filter attn})^2 \qquad \text{at baseband} \qquad (6\text{-}8)$$

$$= \sum_{0}^{f_{\text{hi}}/f_s} (0.1 \, V_{\text{FS}})^2 \left[\frac{1}{\sqrt{1 + \left(\dfrac{f_s}{f_c}\right)^{2n}}} \right]^2$$

$$= \sum_{0}^{1} (0.01 \, V_{\text{FS}}^2) \left[\frac{1}{\sqrt{1 + \left(\dfrac{24}{3}\right)^6}} \right]^2$$

$$= 0.038 \times 10^{-6} \cdot V_{\text{FS}}^2 \text{ watt into } 1 \, \Omega$$

$$SNR_{\text{random alias}} = \frac{V_s^2 \text{ rms}}{N_{\text{alias}}} \qquad\qquad (6\text{-}9)$$

$$\varepsilon_{\text{random alias}} = \frac{\sqrt{2} \cdot 100\%}{\sqrt{SNR_{\text{random alias}}}} \qquad\qquad (6\text{-}10)$$

$$= \frac{\sqrt{2} \cdot 100\%}{\sqrt{V_{\text{FS}}^2 / 0.038 \times 10^{-6} V_{\text{FS}}^2}}$$

$$= 0.027\% \text{FS with presampling filter}$$

Coherent alias frequencies capable of interfering with baseband signals are defined by equation (6-6). The amplitude of the aliasing error components expressed as a percent of full scale are provided for both NRZ and RZ sampling by equation (6-7) with the appropriate sinc function argument. Note that this equation may be evaluated to determine the aliasing amplitude error with or without presampling filtering and its effect on aliasing attenuation. For example, consider a 1 Hz signal

BW for a NRZ sampled-data system with an f_s of 24 Hz. A 23 Hz coherent interfering input signal of –6 dB amplitude (50%FS) will be heterodyned both to 1 Hz and 47 Hz by this 24 Hz sampling frequency, with negligible sinc attenuation at 1 Hz and approximately –30 dB at 47 Hz, for a coherent aliasing baseband aliasing error of 50%FS, applying equation (6-7) in the absence of a presampling filter. This is illustrated by Figure 6-5. The addition of a lowpass three-pole ($n = 3$) Butterworth presampling filter with a 3 Hz cutoff frequency, to minimize filter error to 0.1%FS over the signal BW, then provides –52 dB input attenuation to the 23 Hz interfering signal for a negligible 0.12%FS baseband aliasing error shown by the calculations accompanying equation (6-7). This filter may be visualized superimposed on Figure 6-5.

A more complex situation is presented in the case of random noise because of its wideband spectral characteristic. This type of interference exhibits a uniform amplitude representing a Gaussian probability distribution. Aliased baseband noise power N_{alias} is determined as the sum of heterodyned noise source bands between $mf_s - BW \leq f_{noise}$. These bands occur at intervals of f_s in frequency, shown in Figure 6-6 up to a –3 dB band-limiting f_{hi}, such as provided by an input amplifier cutoff frequency preceding the sampler, with f_{hi}/f_s total noise source bands contributing. N_{alias} may be evaluated with or without the attenuation provided by a presampling filter in determining baseband random noise aliasing error, which is expressed as an aliasing signal-to-noise ratio in equations (6-9) and (6-10). The small sinc amplitude attenuation encountered at baseband is omitted for simplicity.

Consider a –20 dB (0.1 FS) example V_{noise} rms level extending from dc to an f_{hi} of 1 kHz. Solution of equations (6-8) through (6-10), in the absence of a filter, yields 0.42 volts full-scale squared (watts) into 1 ohm as N_{alias} with an f_s as before of 24 Hz and 42 source bands summed to 1 kHz for a random noise aliasing error of 90%FS. Consideration of the previous 1 Hz signal BW and 3 Hz cutoff, three-pole Butterworth lowpass filter provides –54 dB average attenuation over the first noise source band centered at f_s. Significantly greater filter attenuation is imposed at higher noise frequencies, resulting in negligible contribution from summed noise source bands greater than one to N_{alias}. The presampling filter effectiveness, therefore, is such that the random noise aliasing error is only 0.027%FS.

Table 6-1 offers an efficient coordination of presampling filter specifications employing a conservative criterion of achieving –40 dB input attenuation at f_o in terms of a required f_s/BW ratio that defines the minimum sample rate for preventing noise aliasing. The foregoing coherent and random noise aliasing examples meet these requirements with their f_s/BW ratios of 24 employing the general application three-pole Butterworth presampling filter, whose cutoff frequency f_c of three times signal BW provides only a nominal device error addition while achieving significant antialiasing protection. RC presampling filters are clearly least efficient and appropriate only for dc signals considering their required f_s/BW ratio to obtain –40 dB aliasing attenuation. Six-pole Butterworth presampling filters are most efficient in conserving sample rate while providing equal aliasing attenuation at the cost of greater filter complexity. A three-pole Bessel filter is unparalleled in its linearity to both amplitude and phase for all signal types as an antialiasing fil-

TABLE 6-1. Coordination of Sample Rate, Signal Bandwidth, and Sinc Function with Presampling Filter for Aliasing Attenuation at the Folding Frequency

Presampling Filter Poles				f_s/BW for −40 dB Attenuation at f_o Including −4 dB Sinc and Filter f_c of			Filter $\overline{\varepsilon_{\%FS}}$ per Signal Type	
Application	RC	Bessel	Butterworth	20 BW	10 BW	3 BW	DC, Sines	Harmonic
DC signals	1			2560			0.10	1.20
Linear phase		3			80		0.10	0.10
General			3			24	0.10	0.11
Brickwall			6			12	0.05	0.15

ter, but requires an inefficient f_s/BW ratio to compensate for its passband amplitude rolloff. The following sections consider the effect of sample rate on sampled data accuracy—first as step-interpolated data principally encountered on a computer data bus, and then including postfilter interpolation associated with output signal reconstruction.

6-3   STEP-INTERPOLATED DATA INTERSAMPLE ERROR

The NRZ-sampling step-interpolated data representation of Figure 6-7 denotes the way converted data are handled in digital computers, whereby the present sample is current data until a new sample is acquired. Both intersample and aperture volts, ΔV_{pp} and $\Delta V'_{pp}$, respectively, are derived in this development as time–amplitude relationships to augment this understanding.

In real-time data conversion systems, the sampling process is followed by quantization and encoding, all of which are embodied in the A/D conversion process described by Figure 5-11. Quantization is a measure of the number of discrete ampli-

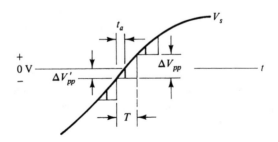

FIGURE 6-7. Intersample and aperture error representation.

tude levels that may be assigned to represent a signal waveform, and is proportional to A/D converter output word length in bits. A/D quantization levels are uniformly spaced between 0 and V_{FS} with each being equal to the LSB interval as described in Figure 5-12. For example, a 12-bit A/D converter provides a quantization interval proportional to 0.024%FS. This typical converter word length thus provides quantization that is sufficiently small to permit intersample error to be evaluated independently without the influence of quantization effects. Note that both intersample and aperture error are system errors, whereas quantization uncertainty is a part of the A/D converter device error.

NRZ sampling is inherent in the operation of S/H, A/D, and D/A devices by virtue of their step-interpolator sampled data representation. Equation (6-11) describes the impulse response for this data representation in the derivation of a frequency domain expression for step interpolator amplitude and phase. Evaluation of the phase term at the sample rate f_s discloses that an NRZ-sampled signal exhibits an average time delay equal to $T/2$ with reference to its input. This linear phase characteristic is illustrated in Figure 6-8. The sampled input signal is acquired as shown in Figure 6-9(a), and represented as discrete amplitude values in analog encoded form. Figure 6-9(b) describes the average signal delay with reference to its input of Figure 6-9(a). The difference between this average signal and its step-interpolator representation in Figure 6-9(b) constitute the peak-to-peak intersample error constructed in Figure 6-9(c).

$$g(t) = U(t) - U(t - T) \tag{6-11}$$

$$g(s) = \frac{1}{s} - \frac{e^{-T}}{s}$$

$$g(j\omega) = \frac{1 - e^{-j\omega T}}{j\omega}$$

$$= T \frac{\sin \pi f T}{\pi f T} \ \angle{-j\omega\,T/2} \qquad \text{NRZ impulse response}$$

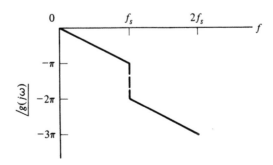

FIGURE 6-8. Step-interpolator phase.

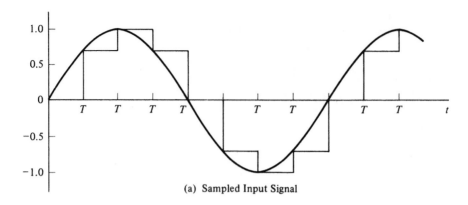

(a) Sampled Input Signal

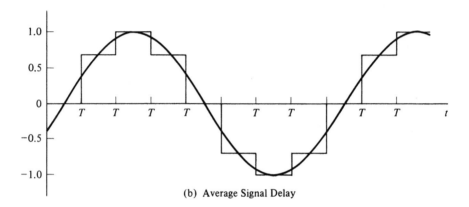

(b) Average Signal Delay

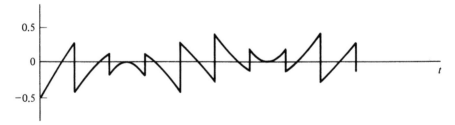

(c) Intersample Error

FIGURE 6-9. Step-interpolator signal representation.

Evaluating delay at $f_s = \dfrac{1}{T}$:

$$\angle\, g(j\omega) = -\pi$$
$$= 2\pi ft$$
$$\therefore t = -\frac{T}{2} \ \text{sec} \qquad \text{sampled signal delay}$$

Equation (6-12) describes the intersample volts ΔV_{pp} for a peak sinusoidal signal V_s evaluated at its maximum rate of change zero crossing shown in Figure 6-7. This representation is converted to ΔV_{rms} through normalization by $2\sqrt{5}$ from the product of the $2\sqrt{2}$ sinusoidal pp–rms factor and the $\sqrt{2.5}$ crest factor triangular step-interpolation contribution of Figure 6-9(c). This expression is also equal to the square root of mean-squared error, which is minimized as a true signal value and its sampled data representation converge. Equation (6-13) reexpresses equation (6-12) to define a more useful amplitude error $\varepsilon_{\Delta V\%FS}$ represented in terms of binary equivalent values in Table 6-2, and is then rearranged in terms of a convenient f_s/BW ratio for application purposes. Describing the signal V_s relative

TABLE 6-2. Step-Interpolated Sampled Data Equivalents

Binary Bits (Accuracy)	Intersample Error $\varepsilon_{\Delta V\%FS}$ (1LSB)	f_s/BW (Numerical)	Applications
0	100.0	2	Nyquist limit
1	50.0		
2	25.0		
3	12.5		
4	6.25	32	Digital toys
5	3.12		
6	1.56		
7	0.78		
8	0.39	512	Video systems
9	0.19		
10	0.097		
11	0.049		
12	0.024	8192	Industrial I/O
13	0.012		
14	0.006		
15	0.003		
16	0.0015	131,072	Instrumentation
17	0.0008		
18	0.0004		
19	0.0002		
20	0.0001	2,097,152	High-end audio

to the specific V_{FS} scaling also permits accommodation of the influence of signal amplitude on the representative rms intersample error of a digitized waveform. Intersample error thus represents the departure of A/D output data from their corresponding continuous input signal values as a consequence of converter sampling, quantizing, and encoding functions including signal bandwidth and amplitude dynamics. For example, a signal V_s of one-half V_{FS} provides only half the intersample error obtained at full V_{FS}, for a constant signal bandwidth and sample rate.

$$\Delta V_{pp} = T \cdot \frac{dV_s}{dt} \qquad \text{intersample volts} \qquad (6\text{-}12)$$

$$= T \cdot \frac{d}{dt} V_s \sin 2\pi \, BWt|_{t=0}$$

$$= 2\pi \, T \, BW \, V_s$$

$$\Delta V_{rms} = \frac{2\pi \, T \, BW \, V_s}{2\sqrt{5}}$$

$$= \sqrt{MSE} \text{ volts}$$

Determining the step-interpolated intersample error of interest is aided by Table 6-2 and equation (6-13). For example, eight-bit binary accuracy requires an f_s/BW ratio of 512, considering its LSB amplitude value of 0.39%FS. This implies sampling a sinusoid uniformly every 0.77 degree, with the waveform peak amplitude scaled to the full-scale value. This obviously has an influence on the design of sampled-data systems and the allocation of their resources to achieve an intersample error of interest. With harmonic signals, the tenth-harmonic amplitude value typically declines to one-tenth that of the fundamental frequency amplitude such that intersample error remains constant between these signal frequencies for arbitrary sample rates. The f_s/BW ratio of two provides an intersample error reference, defining frequency sampling, that is capable of quantifying only signal polarity changes for BW up to $f_s/2$, the Nyquist limit. Unlike digital measurement and control systems in which quantitative amplitude accuracy is of interest, frequency sampling is employed for information that is encoded in terms of signal frequencies as encountered in communications systems and usually involves qualitative interpretation. For example, digital telephone systems often employ seven-bit accuracy, to meet a human sensory error/distortion perception threshold generally taken as 0.7%FS, whose sampling efficiency is increased over that of step-interpolated data by postfilter interpolation, introduced in Section 6-4.

$$\varepsilon_{\Delta V\%FS} = \frac{\Delta V_{rms}}{V_{FS}/\sqrt{2}} \cdot 100\% \qquad \text{intersample error} \qquad (6\text{-}13)$$

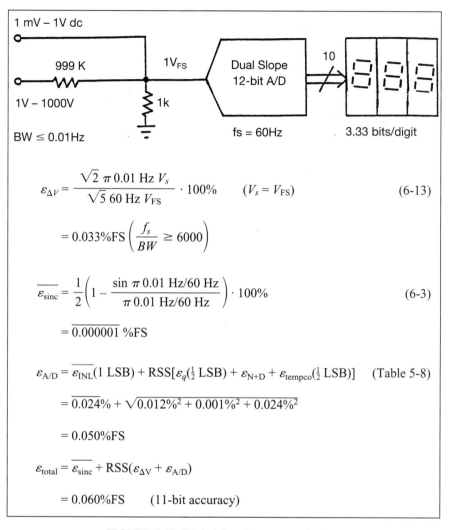

FIGURE 6-10. Digital dc voltmeter error budget.

$$= \frac{\sqrt{2}\pi\, BW\, V_s}{\sqrt{5}\, f_s\, V_{FS}} \cdot 100\%$$

$$\frac{f_s}{BW} = \frac{\sqrt{2}\pi\, V_s\, 100\%}{\sqrt{5}\, \varepsilon_{\Delta V\%FS}\, V_{FS}} \qquad \text{for step-interpolated data}$$

Figure 6-10 describes an elementary digital error budget example of 11-bit binary accuracy for a three-decimal-digit dc digital voltmeter whose 3.33 bits/digit requires 10 bits for display. This acquisition system can accommodate a signal bandwidth to 10 mHz at a sample rate of 60 Hz for an f_s/BW of 6,000. From Chapter 5, intrinsic noise rejection of the integrating A/D converter beneficially provides amplitude nulls to possible voltmeter interference at the f_s value of 60 Hz and –20 dB/decade rolloff to other input frequencies.

Aperture time t_a describes the finite amplitude uncertainty $\Delta V'_{pp}$ within which a sampled signal is acquired, as by a S/H device, referencing Figure 6-7 and equation (6-14), that involves the same relationships expressed in equation (6-12). Otherwise, sampling must be accomplished by a device whose performance is not affected by input signal change during acquisition, such as an integrating A/D. In that direct conversion case, t_a identically becomes the sampling period T. A principal consequence of aperture time is the superposition of an additional sinc function on the sampled-data spectrum. The mean aperture error over the baseband signal described by equation (6-15), however, is independent of the mean sinc error defined by equation (6-3). Although intersample and aperture performance are similar in their relationships, variation in t_a has no influence on intersample error. For example, a fast S/H preceding an A/D converter can provide a small aperture uncertainty, but intersample error continues to be determined by the sampling period T. Figure 6-11 is a nomograph of equation (6-14) that describes aperture error in terms of binary accuracy. Aperture error is negligible in most data conversion systems and consequently not included in the error summary.

$$\Delta V'_{pp} = 2\pi t_a BW\, V_s \qquad \text{aperture volts} \qquad (6\text{-}14)$$

$$\overline{\varepsilon_{a\%FS}} = 1/2\left(1 - \frac{\sin \pi\, BWt_a}{\pi\, BWt_a}\right) \cdot 100\% \qquad (6\text{-}15)$$

6-4 OUTPUT SIGNAL INTERPOLATION, OVERSAMPLING, AND DIGITAL CONDITIONING

The recovery of continuous analog signals from discrete digital signals is required in the majority of instrumentation applications. Signal reconstruction may be viewed from either time domain or frequency domain perspectives. In time domain terms, recovery is similar to interpolation techniques in numerical analysis involving the generation of a locus that reconstructs a signal by connecting discrete data samples. In the frequency domain, efficient signal recovery involves band-limiting a D/A output with a lowpass postfilter to attenuate image spectra present above the baseband signal. It is of further interest to pursue signal reconstruction methods that are more efficient in sample rate requirements than the step-interpolator signal representation described in Table 6-2.

Figure 6-12 illustrates direct-D/A signal recovery with extensions that add both linear interpolator and postfilter functions. Signal delay is problematic in digital control systems such that a direct-D/A output is employed with image spectra attenuation achieved by the associated process closed-loop bandwidth. This method is evaluated in Chapter 7, Section 7-1. Linear interpolation is a capable reconstruction

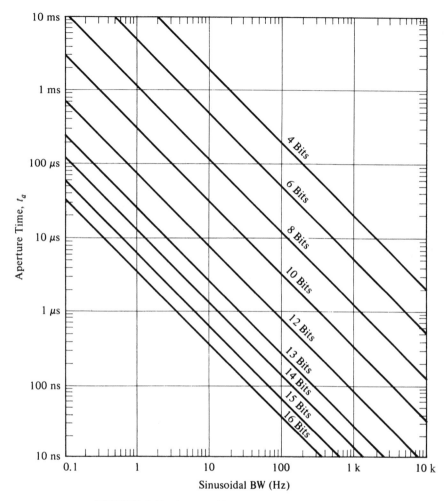

FIGURE 6-11. Aperture binary accuracy nomograph.

function, but achieving a nominal recovery device error is problematical. Linear interpolator effectiveness is defined by first-order polynomials whose line segment slopes describe the difference between consecutive data samples.

Figure 6-13 shows a frequency domain representation of a sampled signal of bandwidth BW with images about the sampling frequency f_s. This ensemble illustrates image spectra attenuated by the sinc function and lowpass postfilter in achieving convergence of the total sampled data ensemble to its ideal baseband BW value. An infinite-series expression of the image spectra summation is given by equation (6-16) that equals the mean squared error (MSE) for direct-D/A output. It follows that step-interpolated signal intersample error may be evaluated by equation (6-17), employing this MSE in defining the D/A output interpolator function of

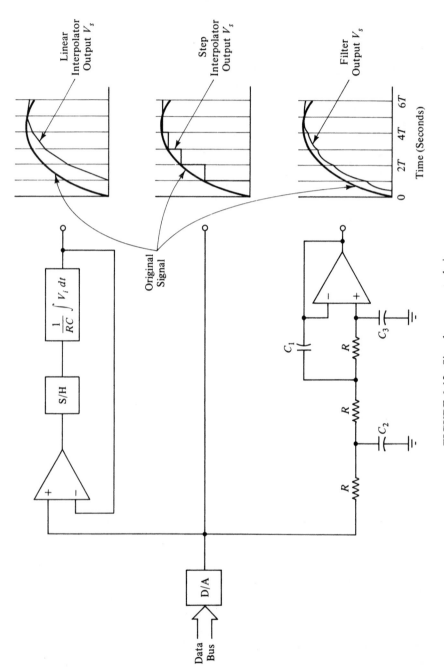

FIGURE 6-12. Signal recovery techniques.

138

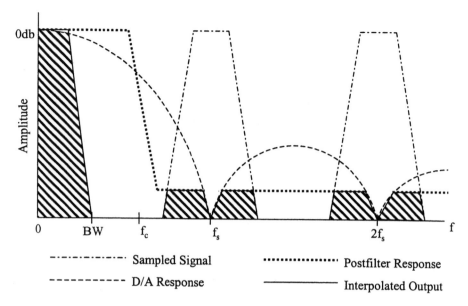

- - - - - - - - Sampled Signal   • • • • • • • • • • • Postfilter Response

- - - - - - - - D/A Response   ——————— Interpolated Output

FIGURE 6-13. Signal recovery spectral ensemble.

Table 6-3, whose result corresponds identically to that of equations (6-12) and (6-13). Note that the sinc terms of equation (6-17) are evaluated at the worst-case first image maximum amplitude frequencies of $f_s \pm BW$.

$$MSE = V_s^2 \sum_{k=1}^{x} \left[\text{sinc}^2 \left(k - \frac{BW}{f_s} \right) + \text{sinc}^2 \left(k + \frac{BW}{f_s} \right) \right] \quad \text{D/A output} \qquad (6\text{-}16)$$

$$= 2V_s^2 \left[\text{sinc}^2 \left(1 - \frac{BW}{f_s} \right) + \text{sinc}^2 \left(1 + \frac{BW}{f_s} \right) \right]$$

$$\varepsilon_{\Delta V\%\text{FS}} = \left[\frac{V_{o\text{FS}}^2}{2V_s^2 \left[\text{sinc}^2 \left(1 - \frac{BW}{f_s} \right) + \text{sinc}^2 \left(1 + \frac{BW}{f_s} \right) \right]} \right]^{-1/2} \cdot 100\% \quad (6\text{-}17)$$

The choice of interpolator function should include a comparison of realizable signal intersample error and the error addition provided by the interpolator device with the goal of realizing not greater than parity in these values. Figure 6-14 shows a comparison of four output interpolators for an example sinusoidal signal at a modest f_s/BW ratio of 10. The three-pole Butterworth posifilter is especially versatile for image spectra attenuation with dc, sinusoidal, and harmonic signals and adds only nominal device error (see Tables 3-5 and 3-6). Its six-bit improvement over direct-D/A recovery is substantial with significant convergence toward ideal signal recon-

TABLE 6-3. Output Interpolator Functions

Interpolater	Amplitude	Intersample Error, $\varepsilon_{\Delta V\%\text{FS}}$
D/A	$\text{sinc}\,(f/f_s)$	$\left[\dfrac{V_{o_{\text{FS}}}^2}{2V_s^2\left[\text{sinc}^2\left(1-\dfrac{BW}{f_s}\right)+\text{sinc}^2\left(1+\dfrac{BW}{f_s}\right)\right]}\right]^{-1/2}\cdot 100\%$
D/A + linear	$\text{sinc}^2\,(f/f_s)$	$\left[\dfrac{V_{o_{\text{FS}}}^2}{V_s^2\left[\text{sinc}^4\left(1-\dfrac{BW}{f_s}\right)+\text{sinc}^4\left(1+\dfrac{BW}{f_s}\right)\right]}\right]^{-1/2}\cdot 100\%$
D/A + one-pole RC	$\text{sinc}\,(f/f_s)[1+(f/f_c)^2]^{-1/2}$	$\left[\dfrac{V_{o_{\text{FS}}}^2}{V_s^2\left[\text{sinc}^2\left(1-\dfrac{BW}{f_s}\right)\left[1+\left(\dfrac{f_s-BW}{f_c}\right)^{2n}\right]^{-1}+\text{sinc}^2\left(1+\dfrac{BW}{f_s}\right)\left[1+\left(\dfrac{f_s+BW}{f_c}\right)^{2n}\right]^{-1}\right]}\right]^{-1/2}\cdot 100\%$
D/A + Butterworth n-pole lowpass	$\text{sinc}\,(f/f_s)[1+(f/f_c)^{2n}]^{-1/2}$	$f_s \pm BW$ substituted for f

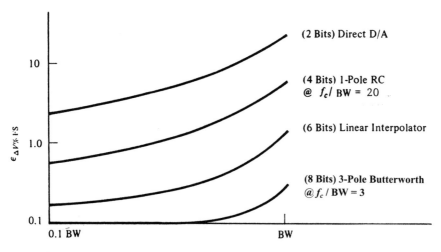

FIGURE 6-14. Output interpolator comparison (sinusoid, $f_s/BW = 10$, $V_s = V_{FS}$).

struction. Bessel filters require an f_c/BW of 20 to obtain a nominal device error, which limits their effectiveness to that of a one-pole RC. However, this does not diminish the utility of Bessel filters for reconstructing phase signals. Interpolator residual intersample error values revealed at fractional BW in Figure 6-14 also define the minimum data word lengths necessary to preserve the interpolated signal accuracy achieved. Table 6-4 describes the signal time delay encountered in transit through the respective interpolation functions.

The consideration of digital signal conditioning prior to output signal recovery is described for both oversampled and interpolative D/A conversion. Oversampled data conversion, introduced by the sigma–delta A/D converter of Figure 5-20, relies upon the increased quantization SNR of 6 dB for each fourfold increase in f_s enabling one binary bit equivalent of additional resolution from Table 5-7. The merit of oversampled D/A conversion, compared to Nyquist sampling and recovery, in

TABLE 6-4. Interpolation Transfer Delay

Interpolator	Time (seconds)
D/A	$\dfrac{1}{2f_s}$
D/A + linear	$\dfrac{1}{2f_s} + \dfrac{1}{f_s}$
D/A + Butterworth n-pole	$\dfrac{1}{2f_s} + \dfrac{n}{4f_c}$

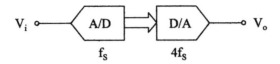

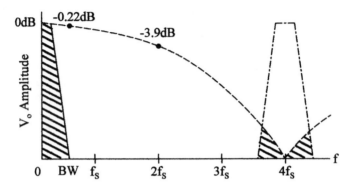

FIGURE 6-15. Oversampled D/A conversion spectrum.

which signal *BW* may exist up to the folding frequency value of $f_s/2$, is a comparable output SNR improvement without increasing the converter word length accompanied by reduced sinc attenuation with reference to equation (6-3). Figure 6-15 shows the performance improvement of four times oversampling D/A conversion with the sampled signal present every fourth sample. First, the fixed quantization noise power for any D/A word length is now distributed over four times the spectral occupancy, such that only one-fourth of this noise is in the signal bandpass to *BW*, yielding a 6 dB SNR improvement. Further, the accompanying sinc amplitude attenuation at signal *BW* is –0.22 dB compared to –3.9 dB encountered with Nyquist

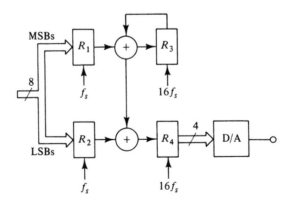

FIGURE 6-16. Interpolative D/A conversion.

TABLE 6-5. Binary Accumulation Assembler Code

```
            SUB    A          ; SET A TO 0
            OUT    20H        ; MAKE PORT 21H ALL INPUTS
            STA    20FFH
            MVI    A,0FFH     ; MAKE PORT 1 ALL OUTPUTS
            OUT    03H
            ANI    04H
            OUT    02H        ; MAKE PORT 0 ALL INPUTS EXCEPT BIT 3
                   00H        ; SET LOAD LINE HIGH
RESET:
            MVI    E, 08H     ; INITIALIZE ADDEND TO 8 (2EXP (M-1))
            MVI    H, 10H     ; INITIALIZE OUTPUT COUNTER (2EXP (N))
INPUT:
            IN     0
            ANI    0IH
            JZ     INPUT      ; WAIT FOR A/D READY
            IN     21H        ; INPUT 8 BIT WORD FROM A/D
            MOV    D, A       ; SAVE IN D
            ANI    0F0H       ; MASK OUT THE 4 LSB'S
            MOV    B, A       ; SAVE MSB'S IN B
            MOV    A, D       ; GET INPUT 8 BITS AGAIN
            ANI    0FH        ; MASK OUT THE 4 MSB'S
            MOV    C, A       ; SAVE LSB'S IN C
ADDPR:
            MOV    A, E       ; GET THE ADDEND
            ADD    C          ; ADD THE LSB'S
            MOV    D, A       ; SAVE THE ADDITION INCLUDING POSSIBLE CARRY
            ANI    0F0H       ; CONVERT ADDITION MODULO 2EXP(N) = 16
            MOV    E, A       ; SAVE NEW ADDEND
            MOV    A, B       ; GET MSB'S
            ADD    D          ; ADD SUM OF ADDEND AND LSB'S WITH POSSIBLE CY
            ANI    OFOH       ; FORCE LSB'S TO 0
            JNZ    NOVER      ; IF NZ, THERE WAS NO OVERFLOW
            MOV    A, B       ; WAS AN OVERFLOW, RESTORE ORIGINAL MSB'S
NOVER:
            MOV    D, A       ; SAVE DATA TO OUTPUT, MSB'S = DATA, LSB'S = 0
OUTOK:
            IN     0H         ; TEST FOR OUTOK PULSE
            ANI    2H
            JZ     OUTOK
            MOV    A, D       ; GET DATA TO OUTPUT
            OUT    1H
            SUB    A          ; SET A TO 0
            OUT    0          ; SEND LOAD LINE LOW
            ORI    4H         ; SEND LOAD LINE HIGH
            OUT    0
            DCR    H          ; DECREMENT MODULO 16 LOOP COUNTER
            JNZ    ADDPR      ; DO 16 TIMES
            JMP    RESET      ; ELSE GET NEXT INPUT DATA

            END
```

sampled signals. Equivalent oversampling performance additions may be obtained at $16 f_s$, $64 f_s$, and higher multiples. Note that while oversampling reduces quantization noise, it cannot increase the data content of a signal.

The rationale for interpolative D/A conversion is the substitution of digital data algorithmic manipulation for reduced resolution hardware without loss of recovered signal accuracy. Its utility is intended for commercial applications, such as digital telephony, where economies of scale can be realized when hardware costs are significant. Instrumentation applications, however, are better served by conventional signal recovery methods in which the D/A converter and data bus word lengths are matched. Figure 6-16 shows a binary accumulation interpolative D/A converter with eight-bit input data. The four MSB values generate a modulation pattern having a frequency of 2^4 times the input sample rate f_s in register R3. When its sum exceeds $2^4 - 1$, a carry is generated and summed with the four LSB values in register R4. The four-bit D/A hardware converter is accordingly updated at $16 f_s$ with an interpolated fine structure that embodies the complete eight-bit data. Realizing the accuracy inherent in the eight-bit data requires effective postfiltering of the oversampled modulation spectra, which is not shown in Figure 6-16. Table 6-5 describes the binary accumulation algorithm assembler code, and Figure 6-17 the time domain interpolative D/A ouput signal prior to postfiltering.

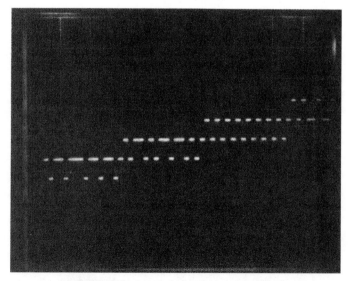

FIGURE 6-17. Interpolative D/A output signal.

BIBLIOGRAPHY

1. W. R. Bennett and J. R. Davey, *Data Transmission,* New York: McGraw-Hill, 1965.

2. W. B. Davenport, Jr. and W. L. Root, *An Introduction to the Theory of Random Signals and Noise,* New York: McGraw-Hill, 1958.

3. L. W. Gardenshire, "Selecting Sample Rates," *ISA Journal,* April 1964.

4. P. H. Garrett, *Analog I/O Design Acquisition: Conversion: Recovery,* Reston, VA: Reston Publishing Co., 1981.

5. A. J. Jerri, "The Shannon Sampling Theorem—Its Various Extensions and Applications: A Tutorial Review," *Proceedings of the IEEE, 65,* 11, November 1977.

6. H. R. Raemer, *Statistical Communication Theory and Applications,* Englewood Cliffs, NJ: Prentice-Hall, 1969.

7. R. W. Lucky, J. Saltz, and E. J. Weldon, *Principles of Data Communication,* New York: McGraw-Hill, 1968.

8. H. Nyquist, "Certain Topics in Telegraph Transmission Theory," *Transactions of the AIEE, 47,* February 1928.

9. A. Papoulis, *Probability, Random Variables, and Stochastic Processes,* New York: McGraw-Hill, 1965.

10. M. Schwartz, W. R. Bennett, and S. Stein, *Communications Systems and Techniques,* New York: McGraw-Hill, 1966.

11. C. E. Shannon and W. Weaver, *The Mathematical Theory of Communication,* Urbana, IL: University of Illinois Press, 1949.

12. E. T. Whittaker, "On Functions which are Represented by the Expansions of the Interpolation Theory," *Proceedings of the Royal Society, 35,* 1915.

13. N. Wiener, *Extrapolation, Interpolation, and Smoothing of Stationary Time Series with Engineering Applications,* Cambridge, MA: MIT Press, 1949.

14. J. M. Wozencraft and I. M. Jacobs, *Principles of Communication Engineering,* New York: Wiley, 1965.

15. A. Kolmogoroff, "Interpolation and Extrapolation von Stationaren Zufalligen Folgen," *Bulletin Academic Sciences, Serial Mathematics, 5,* (USSR), 1941.

16. J. P. Brockman, "An Expert, Error-Referenced Multirate Sampled Data System," Senior Design Thesis, Electrical Engineering Technology, University of Cincinnati, 1984.

7

MEASUREMENT AND
CONTROL INSTRUMENTATION
ERROR ANALYSIS

7-0 INTRODUCTION

Systems engineering considerations increasingly require that real-time I/O systems fully achieve necessary data accuracy without overdesign and its associated costs. In pursuit of those goals, this chapter assembles the error models derived in previous chapters for computer interfacing system functions into a unified instrumentation analysis suite, including the capability for evaluating alternate designs in overall system optimization. This is especially of value in high-performance applications for appraising alternative I/O products.

The following sections describe a low data rate system for a digital controller whose evaluation includes the influence of closed-loop bandwidth on intersample error and on total instrumentation error. Video acquisition is then presented for a high data rate system example showing the relationship between data bandwidth, conversion rate, and display time constant on system performance. Finally, a high-end I/O system example combines premium performance signal conditioning with wide-range data converter devices to demonstrate the end-to-end optimization goal for any system element of not exceeding 0.1%FS error contribution to the total instrumentation error budget.

7-1 LOW-DATA-RATE DIGITAL CONTROL INSTRUMENTATION

International competitiveness has prompted a renewed emphasis on the development of advanced manufacturing processes and associated control systems whose complexity challenge human abilities in their design. It is of interest that conventional PID controllers are beneficially employed in a majority of these systems at the process interface level to obtain industry standard functions useful for integrating process operations, such as control tuning regimes and distributed communications. In fact, for many applications, these controllers are deployed to acquire process measurements, absent control actuation, owing to the utility of their sensor signal conditioning electronics. More significant is an illustration of how control performance is influenced by the controller instrumentation.

Figure 7-1 illustrates a common digital controller instrumentation design. For continuity, the thermocouple signal conditioning example of Figure 4-5 is employed for the controller feedback electronics front end that acquires the sensed process temperature variable T, including determination of its error. Further, the transfer function parameters described by equation (7-1) are for a generic dominant pole thermal process, also shown in Figure 7-1, that can be adapted to other processes as required. When the process time constant τ_0 is known, equation (7-2) can be employed to evaluate the analytically significant closed-loop bandwidth BW_{CL} –3 dB frequency response. Alternately, closed-loop bandwidth may be evaluated experimentally from equation (7-3) by plotting the controlled variable C rise time t_r resulting from setpoint step excitation changes at R.

$$\frac{C}{R} = \frac{K_P K_C \left(1 + \dfrac{1}{2\pi Is} + \dfrac{s}{2\pi D}\right)}{1 + K_P K_C \left(1 + \dfrac{1}{2\pi Is} + \dfrac{s}{2\pi D}\right)} \cdot \left[\frac{\tau_0 s}{1 + K_P K_C \left(1 + \dfrac{1}{2\pi Is} + \dfrac{s}{2\pi D}\right)}\right] \quad (7\text{-}1)$$

$$BW_{CL} = \frac{1 + K_P K_C \left(1 + \dfrac{1}{2\pi Is} + \dfrac{s}{2\pi D}\right)}{2\pi\tau_0} \text{Hz} \quad \text{dominant-pole closed-loop bandwidth} \quad (7\text{-}2)$$

$$BW_{CL} = \frac{2.2}{2\pi t_r} \text{ Hz} \quad \text{universal closed-loop bandwidth} \quad (7\text{-}3)$$

For simplicity of analysis, the product of combined controller, actuator, and process gains K is assumed to approximate unity, common for a conventionally tuned control loop, and an example one-second process time constant enables the choice of an unconditionally stable controller sampling period T of 0.1 sec ($f_s = 10$ Hz) by the development of Figure 7-2. The denominator of the z-transformed transfer function defines the joint influence of K and T on its root solutions, and hence stability within the z-plane unit circle stability boundary. Inverse transformation and evaluation by substitution of the controlled variable $c(n)$ in the time domain analytically reveals a 10–90% amplitude rise time t_r value of 10 sampling periods, or 1 sec, for unit step excitation. Equation (7-3) then approximates a closed-loop bandwidth BW_{CL} value of 0.35 Hz. Table 7-1 provides definitions for symbols employed in this example control system.

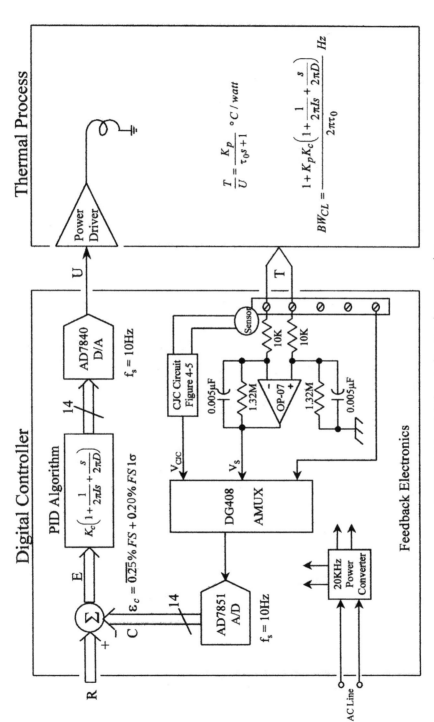

FIGURE 7-1. Digital control system instrumentation.

149

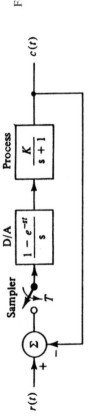
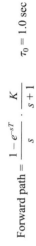

Elementary Sampled Data Control Loop

$$\text{Forward path} = \frac{1 - e^{-sT}}{s} \cdot \frac{K}{s+1} \qquad \tau_0 = 1.0 \text{ sec}$$

$$= K \cdot \frac{(1 - e^{-T})}{(z - e^{-T})} \qquad \text{z-transformed}$$

$$\frac{C(z)}{R(z)} = \frac{\text{Forward path}}{1 + \text{Forward path}} \qquad \text{transfer function}$$

$$= \frac{K(1 - e^{-T})}{z - e^{-T}(1+K) + K}$$

$$C(z) = \frac{K(1 - e^{-T})}{z - e^{-T}(1+K) + K} \cdot \frac{z}{z-1} \qquad \text{unit-step input}$$

$$= \frac{(1 - e^{-0.1})z}{(z - e^{-0.1}(2) + 1)(z-1)} \qquad T = 0.1 \text{ sec}, K = 1.0$$

$$\frac{C(z)}{z} = \frac{(0.1)}{(z - 0.8)(z-1)} \qquad \text{partial fraction expansion}$$

$$= \frac{A}{z - 0.8} + \frac{B}{z-1}$$

$$C(z) = \frac{-0.5\,z}{(z-0.8)} + \frac{0.5\,z}{(z-1)}$$

$$c(n) = [(-0.5)(0.8)^n + (0.5)(1)^n] \cdot U(n) \qquad \text{inverse transform}$$

$$BW_{CL} = \frac{2.2}{2\pi t_r} = 0.35 \text{ Hz} \qquad t_r = nT = 1.0 \text{ sec}$$

FIGURE 7-2. Closed-loop bandwidth evaluation.

TABLE 7-1. Process Control System Legend

Symbol	Dimension	Comment
R	°C	Controller setpoint input
C	°C	Process controlled variable
E	°C	Controller error signal
K_C	watts/°C	Controller proportional gain
I	sec	Controller integral time
D	sec	Controller derivative time
U	watts	Controller output actuation
s	rad/sec	Complex variable
K_P	°C/watts	Process gain
τ_0	sec	Process time constant
t_r	sec	Process response rise time
BW_{CL}	Hz	System closed-loop bandwidth
T	°C	Process sensed variable
V_{CJC}	mV/°C	Cold junction compensation
V_{OFS}	4.096 V_{pk}	Full-scale process variable value
V_s	volts	Process variable signal value

Examination of Figure 7-1 reveals Analog Devices linear and digital conversion components with significant common-mode interference attenuation associated with the signal conditioning amplifier demonstrated in Figure 4-5. The corollary presence of 40 mV of 20 KHz power converter noise at an analog multiplexer input is also shown to result in negligible crosstalk interference as coherent noise sampled data aliasing. A significant result is the influence of the closed-loop bandwidth BW_{CL} on interpolating the controller D/A output by attenuating its sampled data, image frequency spectra. Owing to the dynamics of parameters included in this interpolation operation, intersample error is the dominant contribution to total instrumentation error shown Table 7-2. The 0.45%FS 1σ total controller error approximates eight-bit accuracy, consisting of a $\overline{0.25}$%FS static mean component plus 0.20%FS RSS uncertainty.

Error magnitude declines with reduced electronic device temperatures and less than full-scale signal amplitude V_s encountered at steady-state, as described by the included error models. Largest individual error contributions are attributable to the differential-lag signal conditioning filter and controller D/A-output interpolation. It is notable that the total instrumentation error ε_C value defines the residual variability between the true temperature and the measured controlled variable C, including when C has achieved equality with the setpoint R, and this error cannot further be reduced by skill in controller tuning.

Tuning methods are described in Figure 7-3 that ensure stability and robustness to disturbances by jointly involving process and controller dynamics on-line. Controller gain tuning adjustment outcomes generally result in a total loop gain of approximately unity when the process gain is included. The integrator equivalent value I provides increased gain near 0 Hz to obtain zero steady-state error for the

TABLE 7-2. Digital Control Instrumentation Error Summary

Element	$\varepsilon_{\%FS}$	Comment
Sensor	0.011	Linearized thermocouple (Table 4-5)
Interface	0.032	CJC sensor (Table 4-5)
Amplifier	0.103	OP-07A (Table 4-4)
Filter	0.100	Signal conditioning (Table 3-5)
Signal Quality	0.009	60 Hz ε_{coh} (Table 4-5)
Multiplexer	0.011	Average transfer error
A/D	0.020	14-bit successive approximation
D/A	0.016	14-bit actuation output
Noise aliasing	0.000049	−85 dB AMUX crosstalk from 40 mV @ 20 kHz
Sinc	0.100	Average attenuation over BW_{CL}
Intersample	0.174	Interpolated by BW_{CL} from process τ_0
	0.254%FS	$\overline{\Sigma\text{mean}}$
ε_C	0.204%FS	1σ RSS
	0.458%FS	$\overline{\Sigma\text{mean}} + 1\sigma$ RSS
	1.478%FS	$\overline{\Sigma\text{mean}} + 6\sigma$ RSS

controlled variable C. This effectively furnishes a control loop passband for accommodating the bandwidth of the error signal E. The lead element derivative time D value enhances the transient response for both set point and process load changes to achieve reduced time required for C to equal R.

Analog Multiplexer		
Transfer error		0.01%
Leakage		0.001
Crosstalk		0.00005
ε_{AMUX}	$\overline{\Sigma\text{mean}} + 1\sigma$ RSS	0.011%FS

14-Bit A/D		
Mean integral nonlinearity (1 LSB)		0.006%
Noise + distortion (−80 dB)		0.010
Quantizing uncertainty ($\frac{1}{2}$ LSB)		0.003
Temperature Coefficients ($\frac{1}{2}$ LSB)		0.003
$\varepsilon_{A/D}$	$\overline{\Sigma\text{mean}} + 1\sigma$ RSS	0.020%FS

14-Bit D/A		
Mean integral nonlinearity (1 LSB)		0.006%
Noise + distortion (−80 dB)		0.010
Temperature coefficients ($\frac{1}{2}$ LSB)		0.003
$\varepsilon_{D/A}$	$\overline{\Sigma\text{mean}} + 1\sigma$ RSS	0.016%FS

Noise Aliasing

$$\varepsilon_{\text{coherent alias}} = \text{Interference} \cdot \text{AMUX crosstalk} \cdot \text{sinc} \cdot 100\%$$

$$= \frac{V_{\text{coh}}}{V_{o\text{FS}}} \cdot -85 \text{ dB} \cdot \text{sinc}\left(\frac{mf_s - f_{\text{coh}}}{f_s}\right) \cdot 100\% \quad m \text{ defined at } f_{\text{coh}}$$

$$= \frac{40 \text{ mV}}{4096 \text{ mV}} \cdot (0.00005) \cdot \text{sinc}\left(\frac{2000 \cdot 10 \text{ Hz} - 20 \text{ kHz}}{10 \text{ Hz}}\right) \cdot 100\%$$

$$= 0.000049\%\text{FS}$$

Sinc

$$\varepsilon_{\text{sinc}} = \frac{1}{2}\left(1 - \frac{\sin \pi BW_{\text{CL}}/f_s}{\pi BW_{\text{CL}}/f_s}\right) \cdot 100\%$$

$$= \frac{1}{2}\left(1 - \frac{\sin \pi \, 0.35 \text{ Hz}/10 \text{ Hz}}{\pi \, 0.35 \text{ Hz}/10 \text{ Hz}}\right) \cdot 100\%$$

$$= \overline{0.100}\%\text{FS}$$

Controlled Variable Interpolation

$$\varepsilon_{\Delta V} = \left[\frac{V_{\dot{O}\text{FS}}^2}{V_{\dot{S}}^2 \cdot \left\{\text{sinc}^2\left(1 - \frac{BW_{\text{CL}}}{f_s}\right) \cdot \left[1 + \left(\frac{f_s - BW_{\text{CL}}}{BW_{\text{CL}}}\right)^2\right]^{-1} + \text{sinc}^2\left(1 + \frac{BW_{\text{CL}}}{f_s}\right) \cdot \left[1 + \left(\frac{f_s + BW_{\text{CL}}}{BW_{\text{CL}}}\right)^2\right]^{-1}\right\}}\right]^{-1/2} \cdot 100\%$$

$$= \left[\frac{4.096 \, V^2}{(4.096 \, V)^2 \cdot \left\{\left[\frac{\sin \pi\left(1 - \frac{0.35 \text{ Hz}}{10 \text{ Hz}}\right)}{\pi\left(1 - \frac{0.35 \text{ Hz}}{10 \text{ Hz}}\right)}\right]^2 \cdot \left[1 + \left(\frac{10 \text{ Hz} - 0.35 \text{ Hz}}{0.35 \text{ Hz}}\right)^2\right]^{-1} + \left[\frac{\sin \pi\left(1 + \frac{0.35 \text{ Hz}}{10 \text{ Hz}}\right)}{\pi\left(\frac{1 + 0.35 \text{ Hz}}{10 \text{ Hz}}\right)}\right]^2 \cdot \left[1 + \left(\frac{10 \text{ Hz} + 0.35 \text{ Hz}}{0.35 \text{ Hz}}\right)^2\right]^{-1}\right\}}\right]^{-1/2} \cdot 100\%$$

$$= \left[\cfrac{1}{\left(\cfrac{0.110}{3.03}\right)^2 \cdot (0.001313) + \left(\cfrac{-0.1094}{3.251}\right)^2 \cdot (0.001142)} \right]^{-1/2} \cdot 100\%$$

$$= \quad 0.174\%\text{FS}$$

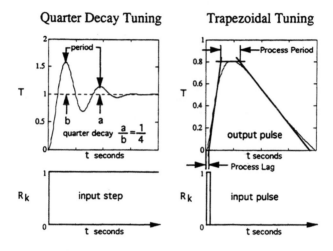

Quarter Decay Tuning   Trapezoidal Tuning

Quarter Decay PID Parameters	Trapezoidal PID Parameters
$P = 1.2 \cfrac{100\%}{\text{Controller } K_c}$ adjusted quarter decay	$P = 100\% \cdot \text{Process Gain}_{\text{trapezoidal tuning}}$
$I = \text{period }_{\text{quarter decay}}, \text{sec}$	$I = \text{Process Period, sec}$
$D = \cfrac{\text{period}}{4} {}_{\text{quarter decay}}, \text{sec}$	$D = 0.44 \,(\text{Process Lag} + \text{Process Period}), \text{sec}$

$$\text{Process Gain}_{\text{trapezoidal tuning}} = \sqrt{\cfrac{\displaystyle\int_{\text{area}} \text{output pulse power} \cdot dt}{\displaystyle\int_{\text{area}} \text{input pulse power} \cdot dt}}$$

FIGURE 7-3. Process controller tuning algorithms.

7-2   HIGH-DATA-RATE VIDEO ACQUISITION

Industrial machine vision, laboratory spectral analysis, and medical imaging instrumentation are all supported by advances in digital signal processing, frequently coupled to television standards and computer graphics technology. Real-time imaging systems usefully employ line-scanned television standards such as RS-343A and RS-170 that generate 30 frames per second, with 525 lines per frame interlaced into one even-line and one odd-line field per frame. Each line has a sweep rate of 53.3 μsec, plus 10.2 sec for the horizontal retrace. The bandwidth required to represent discrete picture elements (pixels) considers the discrimination of active and inactive pixels of equal width in time along a scanning line. The resulting spectrum is defined by Goldman in Figure 7-4, from scan-line timing, as the minimum bandwidth that captures baseband pixel energy [6].

The implementation of a high-speed data conversion system is largely a wideband analog design task. Baseline considerations include employing data converters possessing intrinsic speed with low spurious performance. The example ADS822 A/D converter by Burr-Brown is capable of a 40 megasample per second conversion rate employing a pipelined architecture for input signals up to 10 MHz bandwidth with a 10-bit output word length that limits quantization noise to –60 dB. A one-pole RC input filter with a 15 MHz cutoff frequency is coincident with the conversion-rate folding frequency f_o to provide antialiasing attenuation of wideband input noise.

Figure 7-4 reveals that the performance of this video imaging system is dominated by intersample error that achieves a nominal five-bit binary accuracy, providing 32 luminance levels for each reconstructed pixel. A detailed system error budget, therefore, will not reveal additional influence on performance. The Analog Devices 10-bit ADV7128 pipelined D/A converter with a high-impedance video current output is a compatible data reconstructor providing glitchless performance. Interpolation is achieved by the time constant of the video display for image reconstruction, whose performance is comparable to the response of a single-pole lowpass filter constrained by the 30 frames per second television standard. An efficient microprogrammed input channel containing a high-speed sequencer is also suggested in Figure 7-4 that is capable of executing a complete data-word transfer during each clock cycle to assist in high-data-rate interfacing.

Video Interpolation

$$\varepsilon_{\Delta V} = \left[\frac{V_{0_{FS}}^2}{V_S^2 \cdot \left\{ \mathrm{sinc}^2\!\left(1 - \frac{BW_{\text{pixel}}}{f_s}\right) \cdot \left[1 + \left(\frac{f_s - BW_{\text{pixel}}}{f_{\text{phosphor}}}\right)^2\right]^{-1} + \mathrm{sinc}^2\!\left(1 + \frac{BW_{\text{pixel}}}{f_s}\right) \cdot \left[1 + \left(\frac{f_s + BW_{\text{pixel}}}{f_{\text{phosphor}}}\right)^2\right]^{-1}\right\}} \right]^{-1/2} \cdot 100\%$$

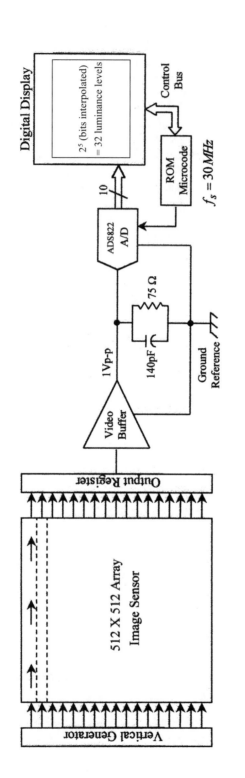

FIGURE 7-4. Video data conversion system.

$$BW_{pixel} = \cfrac{1}{2\left(\cfrac{53.3\mu s \, / \, line}{512 \, pixels \, / \, line}\right)} = 4.8 \, MHz$$

$$f_c = \cfrac{1}{2\pi(140 \, pF)(75\Omega)} = 15 \, MHz$$

$$f_{phospher} = \cfrac{1}{2\pi\left(\cfrac{1}{30} \, sec\right)} = 4.77 \, MHz$$

$$
= \left[\frac{1 V^2}{1 V^2 \cdot \left\{ \left[\frac{\sin \pi\left(1 - \dfrac{4.8\,M}{30\,M}\right)}{\pi\left(1 - \dfrac{4.8\,M}{30\,M}\right)} \right]^2 \cdot \left[1 + \left(\dfrac{30\,M - 4.8\,M}{4.77\,M}\right)^2 \right]^{-1} + \left[\frac{\sin \pi\left(1 + \dfrac{4.8\,M}{30\,M}\right)}{\pi\left(1 + \dfrac{4.8\,M}{30\,M}\right)} \right]^2 \cdot \left[1 + \left(\dfrac{30\,M + 4.8\,M}{4.77\,M}\right)^2 \right]^{-1} \right\}} \right]^{-1/2} \cdot 100\%
$$

$$
= \left[\frac{1}{\left(\dfrac{0.482}{2.636}\right)^2 \cdot (0.034) + \left(\dfrac{-0.482}{3.644}\right)^2 \cdot (0.018)} \right]^{-1/2} \cdot 100\%
$$

$$
= \quad 3.74\%FS \text{ five-bits interpolated video}
$$

7-3   COMPUTER-INTEGRATED INSTRUMENTATION ANALYSIS SUITE

Computer-integrated instrumentation is widely employed to interface analog measurement signals to digital systems. It is common for applications to involve joint input/output operations, in which analog signals are recovered for actuation or end use purposes following digital processing. Instrumentation error models derived for devices and transfer functions in the preceding chapters are presently assembled into an ordered instrumentation analysis suite for I/O system design. This workbook enables evaluating the cumulative error of conditioned and converted sensor signals input to a computer digital data bus, including their output reconstruction in analog form, with the capability for substituting alternate circuit topologies and devices for overall system optimization. This is especially of value for appraising I/O products for implementation selection.

Figure 7-5 describes a high-end I/O system combining the signal conditioning example of Figure 4-6 with the addition of Datel data converter devices to interface a tunable digital bandpass filter for frequency resolution of vibration amplitude signals. Signal conditioning includes a premium performance acquisition channel consisting of a 0.1%FS systematic error piezoresistive bridge strain gauge accelerometer that is biased by isolated ±0.5 V dc regulated excitation and connected differentially to an Analog Devices AD624C preamplifier accompanied by up to 1 V rms of common mode random noise. The harmonic sensor signal has a maximum amplitude of 70 mV rms, corresponding to ±10 g, up to 100 Hz fundamental frequencies with a first-order rolloff to 7 mV rms at a 1 KHz bandwidth. The preampli-

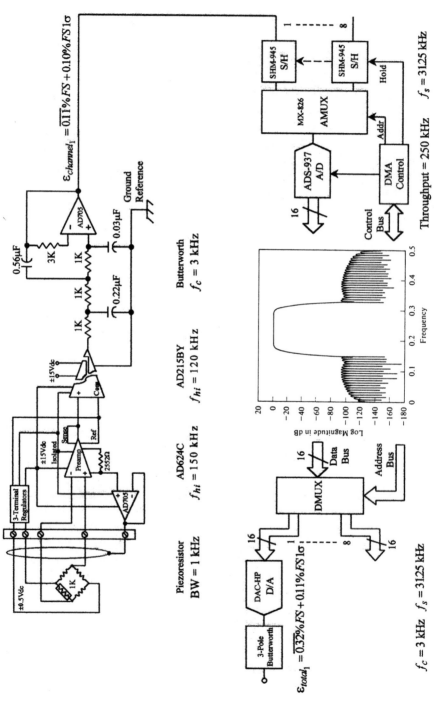

FIGURE 7-5. Vibration analyzer I/O system.

fier differential gain of 50 raises this signal to a ± 5 V_{pp} full-scale value while attenuating the random interference, in concert with the presampling filter, to 0.006%FS signal quality or 212 μV output rms (from ± 5 $V/\sqrt{2}$ rms times 0.00006 numerical). The associated sensor-loop internal noise of 15 μV_{pp} plus preamplifier referred-to-input errors total 27 μV dc with reference to Table 4-4. This defines a signal dynamic range of $\sqrt{2} \cdot 70$ mV/27 μV, or 71 dB, approximating 12 bits of amplitude resolution. Amplitude resolution is not further limited by subsequent system devices that actually exceed this performance, such as the 16-bit data converters.

It is notable that the Butterworth lowpass presampling signal conditioning filter achieves signal quality upgrading for random noise through a linear filter approximation to matched filter efficiency by the provisions of Chapter 4. This filter also coordinates undersampled noise aliasing attenuation described in Chapter 6 with cutoff frequency derating to minimize its mean filter error from Chapter 3. Errors associated with the amplifiers, S/H, AMUX, A/D, and D/A data converters are primarily nonlinearities and temperature drift contributions that result in LSB equivalents between 12–15 bits of accuracy. The A/D and DIA converters are also discrete switching devices to avoid signal artifacts possible with sigma–delta type converters. Sample rate f_s, determined by dividing the available 250 KHz DMA transfer rate between eight channels, is thirty-one times the 1 KHz signal BW, which provides excellent sampled-data performance in terms of small sinc error, negligible noise aliasing of the 212 μV rms of residual random interference by modestly exceeding the minimum f_s/BW ratio of 24 from Table 6-1, and accurate output reconstruction.

Figure 7-6 shows the error of converted input signal versus frequency applied to a digital data bus, where its zero order hold intersample error value is the dominant contributor at 0.63%FS at full bandwidth. The combined total input error of 0.83%FS remains constant from 10% of signal bandwidth to the 1 KHz full bandwidth value, owing to harmonic signal amplitude rolloff with increasing frequency, declining to 0.32%FS at 1% bandwidth. It is significant that the sampled image frequency spectra described in Chapter 6 are regenerated by each I/O sampling operation from S/H through D/A converter devices, and that these spectra are transformed with signal transfer from device to device when there is a change in f_s. Increasing f_s accordingly results both in sampled image frequency spectra being heterodyned to higher frequencies and a decreased mean signal attenuation from the associated sinc function. This describes the basis of oversampling, defined as sampling rates greater than the Nyquist f_s/BW ratio of two in Section 6-4, which offers enhanced output reconstruction through improved attenuation of the higher sampled image frequency spectra by the final postfiltering interpolator.

The illustrated I/O system and its accompanying analysis suite models provide detailed accountability of total system performance and realize the end-to-end optimization goal of not exceeding 0.1%FS error for any contributing element to the error summary of Table 7-3. Output signal reconstruction is effectively performed by a post-D/A Butterworth third-order lowpass filter derated to reduce its component error while simultaneously lowering intersample error. This implementation results in an ideal flat total 1σ instrumentation error versus bandwidth, shown in Figure 7-6, of 0.43%FS. This error is equivalent to approximately eight bits of true amplitude

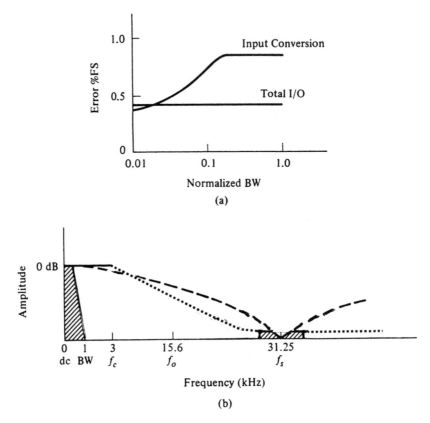

FIGURE 7-6. I/O system total error and spectra.

accuracy within 12 bits of signal dynamic range and 16 bits of data quantization. Six-sigma confidence is defined by the extended value of 0.97%FS, consisting of one mean plus six RSS error values.

The Microsoft Excel spreadsheet contains an interactive workbook of complete instrumentation system error models in 69 Kbytes for computer-assisted engineering design. The first page of this six sheet analysis suite permits defining sensor and excitation input values, including signal bandwidth and differential signal voltage amplitude, and provides for both random and coherent interference. This data is utilized for subsequent model calculations, and returns the input signal-to-noise ratio and required system voltage gain. Examples of values for the vibration analyzer I/O system shown in Figure 7-5 are given throughout the pages of this spreadsheet. Specific sensor and excitation input values and model calculations associated with this example are presented in greater detail in Figure 4-6 and the accompanying text.

The second and third pages accommodate up to four cascaded amplifiers per system, whereby thirteen parameters are entered for each amplifer selected from manu-

TABLE 7-3. I/O Instrumentation Error Summary

Element	$\varepsilon_{\%FS}$	Comment
Sensor	0.100000	Piezoresistor bridge
Interface	0.010000	Residual differential excitation
Amplifiers	0.033950	AD624C + AD215BY
Presampling filter	0.115000	Three-pole Butterworth
Signal quality	0.006023	Random noise ε_{rand}
Noise aliasing	0.000007	212 μV residual interference
Sinc	0.084178	Average signal attenuation
Multiplexer	0.004001	Average transfer error
Sample hold	0.020633	400 ns acquisition time
A/D	0.002442	16-bit subranging
D/A	0.013032	16-bit converter
Interpolator filter	0.1150000	Three-pole Butterworth
Intersample	0.000407	Output interpolation
	0.318179%FS	Σmean
ε_{total}	0.109044%FS	1σ RSS
	0.427223%FS	Σmean + 1σ RSS
	0.972443%FS	Σmean + 6σ RSS

facturer's data. Seven additional quantities related to sensor circuit parameters are also required, which ordinarily accompany only the front-end amplifier in a system. Seven calculated equivalent input error voltages are returned for each amplifier, defining their respective error budgets, and combined in an eighth amplifier value expressing error as %FS. A detailed representation of the model calculations for amplifiers employed in this example are tabulated in Tables 4-3 and 4-4.

The fourth page evaluates linear signal conditioning performance in terms of attained signal quality, including specification of parameters for a band-limiting presampling filter, which serves a dual role in signal conditioning and aliasing prevention. Calculated values returned include residual coherent and random interference error as well as filter device error from Chapter 3. Sampled data parameters including a trial sample rate f_s, and undersampled coherent and random interference amplitude values existing above the Nyquist frequency ($f_s/2$), are then entered so that aliasing error may be evaluated. The amplitude values are proportional to postsignal conditioning residual errors, evaluated by equations (4-15) and (4-16), as determined by scaling to the system full-scale voltage value. Returned values on the fifth spreadsheet page including aliasing error and sinc error rely upon corresponding models developed in Chapter 6.

The remaining analysis suite spreadsheet entries consist of parameter values obtained from manufacturers for modeling five data conversion devices. These include AMUX, S/H, A/D, D/A, and output interpolator devices primarily from Section 6-4. The combined error for all of the device and system contributions are automatically tabulated for full signal BW including output interpolation, and plotted from 1% to 100% of BW both for converted signals on a computer data bus and end-to-end with choice of output interpolator device. Available data reconstructors include direct D/A, one-pole RC, and three-pole Butterworth interpolators.

Computer Integrated Instrumentation Analysis Suite Spreadsheet

Parameter	Symbol	Value	Units	Comment (black: entered; shaded: calculated)
			Sensor and Excitation Entries	
Sensor Error Type	ε	s	M or S	M = Static Mean, S = Variable Systematic
Sensor Error Value	ε_{sensor}	0.1	%FS	Sensor full scale error from manufacturer's information
Peak Input Signal Voltage	V_s	0.1	Volts	Sensor full-scale signal voltage DC or RMS $\sqrt{2}$ up to fundamental = BW/10 for harmonic signals
Peak Output Signal Voltage	V_{oFS}	5.00	Volts	System full-scale voltage DC or RMS $\sqrt{2}$
Signal Bandwidth	BW	1000	Hertz	Sensor signal bandwidth to highest frequency of interest
Interface Error Type	ε	s	M or S	M = Static Mean, S = Variable Systematic
Common Mode Interference	V_{cm}	1.0	Volts	Input common-mode DC or RMS random and/or coherent interference amplitude
Differential Input Signal @ BW	V_{diff}	0.007	Volts	Sensor DC or RMS differential voltage at full BW for signal quality evaluation
Coherent Interference Present	Coh	N	Y or N	Enter N if no coherent interference
Coherent Interference Frequency	f_{coh}	0	Hertz	Frequency of coherent interfering signal if present
Random Interference Present	Rand	y	Y or N	Enter a Y if random noise is present, N If not
Input Interface Error	$\varepsilon_{interface}$	0.01	%FS	Interface termination or sensor excitation error
Sinusoidal/Harmonic	H or S	h		Enter H for complex harmonic signals and S for sinusoidal or DC signals
Required Voltage Gain	A_v	50	V/V	V_{oFS}/V_S total gain between sensor and A/D converter
Input SNR	SNR_i	4.900E-05	$(V/V)^2$	Input signal-to-noise ratio as $(V_{diff}/V_{cm})^2$

Amplifier Data

Parameter	Symbol	Amp_1	Amp_2	Amp_3	Amp_4	Units	Comment
		Amplifier Error Budget Parameters					
Amplifier Type		AD624C	AD215BY				Manufacturer's part number
Common Mode Resistance	R_{cm}	1.00E+09	5.00E+09			Ohms	Input common mode resistance
Differential Resistance	R_{diff}	1.00E+09	1.00E+12			Ohms	Input differential resistance
Amplifier Cutoff Frequency	f_{hi}	150,000	120,000			Hertz	Amplifier −3 dB cutoff frequency
Mean Offset Voltage Amplitude	V_{OS}	2.500E-05	4.000E-04			Volts	DC voltage between differential inputs
Voltage Offset Temperature Drift	$\Delta V_{OS}/\Delta T$	2.50E-07	2.00E-06			Volts/°C	Input offset voltage temperature drift
Temperature Variation	ΔT	10	10			°C	Circuit temperature variation
Offset Current	I_{OS}	0.01	0.3			μA	DC input offset bias current difference
Current Offset Temperature Drift	$\Delta I_{OS}/\Delta t$	2.00E-05	1.00E-03			μA/C°	Input offset current temperature drift
Ambient Temperature	T_{amb}	20	20			°C	Temperature of system environment
Sensor Current Amplitude	I_{DC}	1000	0			μA	Sensor DC current flow if present
Contact Noise Frequency	$f_{contact}$	100	100			Hertz	Contact noise test frequency (convention 10% of BW)
Offset Voltage Nulled	V_{OSNull}	N	N			A or N	Enter A if V_{OS} added to ε_{amp}, N if nulled
Input Noise Voltage Equivalent	V_n	0.004	0			$\mu V/\sqrt{Hz}$	Amplifier RMS noise voltage per root Hertz

		Amplifier Error Budget Parameters					
Parameter	Symbol	Amp_1	Amp_2	Amp_3	Amp_4	Units	Comment
Input Noise Current Equivalent	I_n	0.06	0			pA/$\sqrt{Hz}$	Amplifier RMS noise current per root Hertz
Common Mode Rejection	CMRR	5.00E+05	1.00E+05			V/V	Numeric common-mode rejection ratio
Gain Nonlinearity	$f(A_V)$	10	50			ppm	Gain nonlinearity over gain range
Peak Output Signal Voltage	V_{oFS}	5	5			Volts	Amplitude full-scale output voltage DC or RMS $\cdot \sqrt{2}$
Differential Gain	$A_{V_{diff}}$	50	1			V/V	Closed-loop differential gain
Gain Temperature Drift	$\Delta A_V/\Delta T$	5	15			ppm/C°	Gain temperature drift
Source Resistance	R_s	1000	50			Ohms	Source resistance seen by respective amplifier
Voltage Drift from Temp.	ΔV_{OS}	2.500E-06	2.000E-05	0.000E+00	0000E+00	Volts	Input offset voltage temperature drift
Mean Offset I_{OS} Voltage	$I_{OS}R_s$	1.000E-05	1.500E-05	0.000E+00	0.000E+00	Volts	Voltage error due to input offset current
Thermal Noise	V_t	4.022E-09	8.993E-10	0.000E+00	0.000E+00	V/$\sqrt{Hz}$	Thermal RMS noise in sensor circuit
Contact Noise	V_c	1.802E-09	0.000E+00	0.000E.00	0.000E+00	V/$\sqrt{Hz}$	Contact RMS noise in sensor circuit
Total Noise	V_{Npp}	1.521E-05	2.056E-06	0.000E+00	0.000E+00	Volts	6.6RSS($V_t + V_c + V_n$)$\sqrt{f_{hi}}$
Mean Gain Nonlinearity	$V_{f(A_V)}$	1.000E-06	2.500E-04	0.000E+00	0.000E+00	Volts	Voltage error due to gain nonlinearity
Gain Temperature Drift	$V_{\Delta A_V/\Delta T}$	5.000E-06	7.500E-04	0.000E+00	0.000E+00	Volts	Voltage error due to gain temperature drift
Amplifier Errors	ε_{amp}	0.02721	0.02031	0.00000	0.00000	%FS	(Σ mean V + RSS other V) $\cdot$ ($A_{V_{diff}}/V_{oFS}$) $\cdot$ 100%

Signal Quality and Presampling Filter Entries

Parameter	Symbol	Value	Units	Comment (black: entered; shaded: calculated)
Presampling Filter Poles	n	3	Poles	Valid for 1–8 Butterworth poles for harmonic signals
Filter Present		y	Y or N	Enter N if no filter present
Filter Efficiency	K	0.9	Parameter	Linear filter efficiency relative to matched filtering (0.9 default value)
Mean Filter Error	ε_{filter}	0.115	%FS	Presampling filter error for complex harmonic signal
Filter Cutoff Frequency	f_c	3000	Hertz	Presampling filter cutoff frequency
Amplifier SNR	SNRamp	1.23E+07	W/W	Signal conditioning amplifier output signal-to-noise power ratio
Amplifier SNR Error	$\varepsilon_{amp\ SNR}$	0.02857	%FS	Signal conditioning amplifier output error
Coherent Filter SNR	SNR_{coh}	—	W/W	Filter output coherent signal-to-noise ratio as power ratio
Coherent Filter SNR Error	$\varepsilon_{coh\ amp}$	—	%FS	Filter output full-scale signal error for filter SNR
Random Filter SNR	SNR_{rand}	5.5E+08	W/W	Random filter SNR
Random Filter SNR Error	$\varepsilon_{rand\ amp}$	0.00602	%FS	Filter output full-scale signal error for filter SNR (random)
Total Signal Quality	ε_{sq}	0.00602	%FS	$\varepsilon_{amp\ SNR}$ or RSS ($\varepsilon_{rand\ amp}$ + $\varepsilon_{coh\ amp}$) with filter

Aperture, Sinc, and Aliasing Entries

Parameter	Symbol	Value	Units	Comment (black: entered; shaded: calculated)
Aperture Time	t_a	0.002	μs	Aperture time of sample and hold
Sample Rate	f_s	31250	Hertz	Sample rate selected
Undersampled Coherent Interference	A_{coh}	0	V RMS	Amplitude of the coherent undersampled RMS noise at S/H and A/D
Undersampled Random Interference	A_{rand}	2.12E-04	V RMS	Amplitude of the random undersampled RMS noise at S/H and A/D
Coherent Alias Frequency	$f_{coh\ alias}$	0	Hertz	Undersampled coherent aliasing source frequency at input

Aperture, Sinc, and Aliasing Entries

Parameter	Symbol	Value	Units	Comment (black: entered; shaded: calculated)
Interfering Baseband Alias	f_{alias}	0	Hertz	Baseband coherent aliasing frequency
Mean Aperture error	ε_a	3.290E-10	%FS	Aperture error as percent full scale
ZOH Intersample Error	$\varepsilon_{\Delta VZOH}$	0.6358	%FS	ZOH intersample error at full BW prior to interpolation
Mean Sinc Error	$\varepsilon_{NRZ\ sinc}$	0.0842	%FS	Average sinc error for NRZ sampling
Coherent Alias Error	$\varepsilon_{coh\ alias}$	0.00E+00	%FS	Aliasing by the undersampled coherent interference amplitude
Random Alias Error	$\varepsilon_{rand\ alias}$	7.50E-06	%FS	Aliasing by the undersampled random interference amplitude
Total Alias Error	ε_{alias}	7.50E-06	%FS	RSS($\varepsilon_{coh\ alias}$ + $\varepsilon_{rand\ alias}$)

Multiplexer Entries

Parameter	Symbol	Value	Units	Comment (black: entered; shaded: calculated)
Mean Transfer Error	ε_{trans}	0.003	%FS	Mean transfer error as percent full scale
Crosstalk	ε_{cross}	0.00005	%FS	Crosstalk error as percent full scale
Leakage	ε_{leak}	0.001	%FS	Leakage error as percent full scale
Mean Multiplexer Error	ε_{AMUX}	0.00400	%FS	

Sample-Hold Entries

Parameter	Symbol	Value	Units	Comment (black: entered; shaded: calculated)
Acquisition Error	ε_{acq}	0.00076	%FS	Acquisition error following required settling time
Nonlinearity	ε_{lin}	0.0004	%FS	Sample-hold nonlinearity errors
Gain	ε_{gain}	0.02	%FS	Gain errors
Tempco	ε_{tempco}	0.005	%FS	Temperature coefficient errors
Sample-Hold Error	$\varepsilon_{S/H}$	0.02063	%FS	RSS sample hold entries

Analog-to-Digital Converter Entries

Parameter	Symbol	Value	Units	Comment (black: entered; shaded: calculated)
A/D Wordlength	Data Bus	16	Bits	Converter wordlength
Quantizing Uncertainty	ε_q	0.0008	%FS	Quantizing uncertainty as %FS of 1/2 LSB
Mean Integral Nonlinearity	ε_{INL}	0.0011	%FS	Mean integral nonlinearity as %FS
Noise + Distortion	ε_{N+D}	0.0001	%FS	Noise plus distortion as a %FS
Tempco	ε_{tempco}	0.0011	%FS	Temperature coefficient errors
A/D Error	$\varepsilon_{A/D}$	0.00244	%FS	ε_{INL} + RSS other

Digital-to-Analog Converter Entries

Parameter	Symbol	Value	Units	Comment (black: entered; shaded: calculated)
D/A Wordlength	Data	16	Bits	Converter wordlength
Mean Integral Nonlinearity	ε_{INL}	0.003	%FS	Mean integral nonlinearity as %FS
Tempco	ε_{tempco}	0.01	%FS	Temperature coefficient errors
Noise + Distortion	ε_{N+D}	0.0008	%FS	Noise plus distortion as a %FS
D/A Error	$\varepsilon_{D/A}$	0.01303	%FS	ε_{INL} + RSS other

Data Reconstruction Entries

Parameter	Symbol	Value	Units	Comment (black: entered; shaded: calculated)
Reconstructor Type	Recovery	b	D, R, or B	Type of reconstruction circuit (D = direct D/A, R = 1-pole RC, B = Butterworth 3-pole filter)
Interpolator Poles	N	3	Poles	0 for direct D/A, 1 for 1-pole RC, 3 for Butterworth
Interpolated Intersample Error	$\varepsilon_{\Delta V}$	0.0004	%FS	Interpolated intersample error
Mean Interpolator Device Error	ε_{interp}	0.1150	%FS	Interpolator device error
Reconstruction System Error	ε_{recov}	0.1280	%FS	ε_{interp} + RSS($\varepsilon_{D/A}$ + $\varepsilon_{\Delta V}$)
Cutoff Frequency	f_{cr}	3000	Hertz	Interpolator device filter cutoff frequency

BIBLIOGRAPHY

1. P. H. Garrett, *Analog I/O Design Acquisition: Conversion: Recovery,* Reston, VA: Reston, 1981.
2. E. L. Zuch, *Data Acquisition and Conversion Handbook,* Mansfield, MA: Datel-Intersil, 1979.
3. B. M . Gordon, *The Analogic Data-Conversion Systems Digest,* Wakefield, MA, Analogic, 1977.
4. *Analog-Digital Conversion Handbook,* Norwood, MA, Analog Devices Corp., 1972.
5. I. Bazovsky, *Reliability Theory and Practice,* Englewood Cliffs, NJ: Prentice-Hall, 1965.
6. S. Goldman, *Frequency Analysis, Modulation and Noise,* New York: McGraw-Hill, 1948.
7. U. Peled, "A Design Method With Application to Prefilters and Sampling Rate Selection in Digital Flight Control Systems," Department of Aeronautics and Astronautics, Stanford University, May 1978.
8. P. Katz, Digital Control Using Microprocessors, Englewood Cliffs, NJ: Prentice-Hall, 1981.
9. J. P. Brockman, "An Expert Error-Referenced Multirate Sampled-Data System," Senior Design Thesis, Electrical Engineering Technology, University of Cincinnati, 1984.
10. K. J. Astrom and B. Wittenmark, *Computer Controlled Systems,* Englewood Cliffs, NJ: Prentice-Hall, 1984.
11. P. Wintz and R. C. Gonzales, Digital Image Processing, Reading, MA: Addison-Wesley, 1977.
12. W. Kester, "Test Video A/D Converters Under Dynamic Conditions," *Electronic Design News,* August 18, 1982.
13. M. D. Mesarovic, *Views on General Systems Theory,* New York: Wiley, 1964.
14. J. Sherwin, "Simplify Analog/Computer Interfacing," *Electronic Design,* August 16, 1977.
15. P. H. Garrett, "Optimize Transducer/Computer Interfaces," *Electronic Design,* May 24, 1977.
16. R. L. Morrison, "Getting Transducers to Talk to Digital Computers," *Instruments and Control Systems,* January 1978.
17. E. L. Zuch, "Principles of Data Acquisition and Conversion," *Digital Design,* May 1979.
18. B. M. Gordon, "Digital Sampling and Recovery of Analog Signals," *Electronic Equipment Engineering,* May 1970.
19. L. Solomon and E. Ross, "Educating Dumb Data Acquisition Subsystems," *Digital Design,* November 1976.
20. D. Stantucci, "Maneuvering for Top Speed and High Accuracy in Data Acquisition," *Electronics,* November 27, 1975.

8

MULTISENSOR ARCHITECTURES AND ERROR PROPAGATION

8-0 INTRODUCTION

The purpose of this chapter is to extend the data acquisition error analysis of the preceding chapters to provide understanding about how errors originating in multi-sensor architectures combine and propagate in algorithmic computations. This development is focused on the wider applications of sensor integration for improving data characterization rather than the narrower applications of sensor fusion employed for data ambiguity reduction.

Three diverse multisensor instrumentation architectures are analyzed to explore error propagation influences. These include: sequential multiple sensor information acquired at different times; homogeneous information acquired by multiple sensors related to a common description; and heterogeneous multiple sensing of different information that jointly describe specific features. These architectures are illustrated, respectively, by multisensor examples of airflow measurement through turbine engine blades, large electric machine temperature modeling, and in situ material measurements in advanced process control. Instructive outcomes include the finding that mean error values aggregate with successive algorithmic propagation whose remedy requires minimal inclusion.

8-1 MULTISENSOR FUSION, INTEGRATION, AND ERROR

The preceding chapters have demonstrated comprehensive end-to-end modeling of instrumentation systems from sensor data acquisition through signal conditioning and data conversion functions and, where appropriate, output signal reconstruction and actuation. These system models beneficially provide a physical description of instrumentation performance with regard to device and system choices to verify ful-fillment of measurement accuracy, defined as the complement of error. Total instrumentation error is expressed as a sum of static mean error contributions plus the one sigma root-sum-square (RSS) of systematic and random error variances as a percent of full-scale amplitude. This is utilized throughout the text as a unified measurement instrumentation uncertainty description. Its components are illustrated in Figure 1-1, applicable to each system element beginning with the error of a sensor relative to its true measurand, and proceeding with all inclusive device and instrumentation system error contributions.

Chapter 4, Section 4-4 reveals that combining parallel–redundant instrumentation systems serves to reduce only the systematic contributions to total error through averaging, whereas mean error contributions increase additively to significantly limit the merit of redundant systems. This result emphasizes that good instrumentation design requires minimization of mean error in the signal path as shown for band-limiting filters in Chapter 3. Conversely, additive interference sources are generally found to be insignificant error contributors because of a combination of methods typically instituted for their attenuation. Modeled instrumentation system error, therefore, valuably permits performance to be quantitatively predicted a priori for measurement confidence and data consistency such as sensed-state process observations. Confidence to six sigma is defined for a system as its static mean error plus six times its RSS 1σ error.

Sensor fusion is primarily limited to medical imaging and target recognition applications. Fusion usually involves the transformation of redundant multisensor data into an equivalent format for ambiguity reduction and measured property retrieval otherwise unavailable from single sensors. Data fusion often extracts multiple image or target parametric attributes, including object position estimates, feature vector associations, and kinematics from sources such as sub-Hz seismometers to GHz radar to Angstrom-wavelength spectrometers. Sonar signal processing, illustrated in Figure 8-1, illustrates the basics of multisensor fusion, whereby a sensor array is followed by signal conditioning and then signal processing subprocesses, concluding in a data fusion display. Sensor fusion systems are computationally intensive, requiring complex algorithms to achieve unambiguous performance, and are burdened by marginal signal quality.

This chapter presents multisensor architectures commonly encountered from industrial automation to laboratory measurement applications. With these multisensor information structures, data are not fused, but instead nonredundantly integrated to achieve better attribution and feature characterization than available from single sensors. Three architectures are described that provide understanding concerning integrated multisensor error propagation, where propagation in algorithmic computations is evaluated employing the relationships defined in Table 8-1. A sequential architecture describes multisensor data acquired in different time intervals, then a homogeneous architecture describes the integration of multiple measurements related to a common description. Finally, a heterogeneous architecture describes nonoverlapping multisensor data that jointly account for specific features. The integration of instrumentation systems is separately presented in Chapter 9.

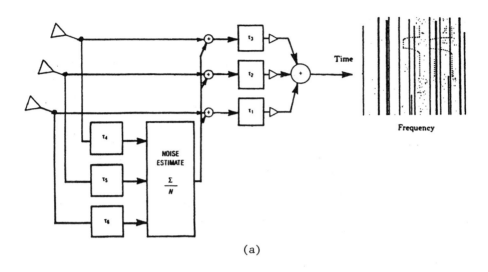

(a)

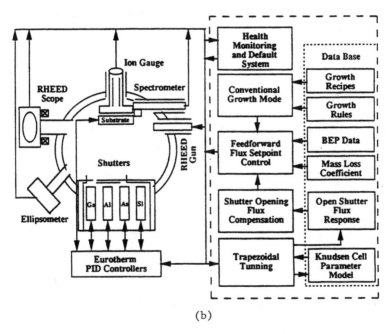

(b)

FIGURE 8-1. (a) Sonar redundant sensor fusion; (b) molecular beam epitaxy nonredundant integration.

TABLE 8-1. Instrumentation Error Algorithmic Propagation

Instrumentation Error	Algorithmic Operation	Error Influence
$\overline{\varepsilon_{mean}}$ %FS	Addition	$\Sigma \varepsilon_{mean}$ %FS
	Subtraction	$\Sigma \varepsilon_{mean}$ %FS
	Multiplication	$\Sigma \varepsilon_{mean}$ %FS
	Division	$\Sigma \varepsilon_{mean}$ %FS
	Power function	$\Sigma \varepsilon_{mean}$ %FS × |exponent value|
ε %FS 1σ	Addition	RSS ε %FS 1σ
	Subtraction	RSS ε %FS 1σ
	Multiplication	RSS ε %FS 1σ
	Division	RSS ε %FS 1σ
	Power function	RSS ε %FS 1σ × |exponent value|

8-2   SEQUENTIAL MULTISENSOR ARCHITECTHRE

Figure 8-2 describes a measurement process applicable to turbine engine manufacture for determining blade internal airflows, with respect to design requirements, essential to part heat transfer and rogue blade screening. A preferred evaluation method is to describe blade airflow in terms of fundamental geometry such as its effective flow area. The implementation of this measurement process is described by analytical equations (8-1) and (8-2), where uncontrolled air density ρ appears as a ratio to effect an air-density-independent airflow measurement. That outcome beneficially enables quantitative determination of part airflows from known parameters and pressure measurements defined in Table 8-2. The airflow process mechanization consists of two plenums with specific volumetric airflows and four pressure measurements.

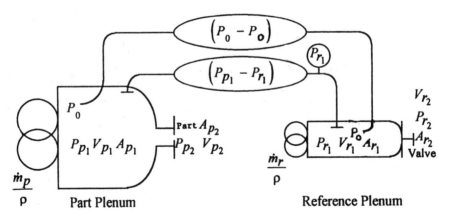

FIGURE 8-2. Multisensor airflow process.

TABLE 8-2. Airflow Process Parameter Glossary

Known Airflow Process Parameters			Measured Airflow Process Parameters		
Symbol	Value	Description	Symbol	Value	Description
$\dot{m}_r$	$\dfrac{ft^3}{min}$	Reference plenum volumetric flow	A_{P2}	ft^2	Part effective flow area
A_{r_1}	ft^2	Reference plenum inlet area	$P_{P_1} - P_{r_1}$	lb/ft^2	Part-to-reference plenum differential pressure
V_{r_1}	$\dfrac{ft}{min}$	Reference plenum inlet velocity	P_{r1}	lb/ft^2	Reference plenum gauge pressure
A_{P_1}	ft^2	Part plenum inlet area	$P_o - P_o$	lb/ft^2	Reference part plenum equalized stagnation pressures
ρ	$0.697E\text{-}6$ $\dfrac{lb - min}{ft^4}$	Air density at standard temperature and pressure	P_{P2}	$P_{atm}\ lb/ft^2$	Part plenum exit pressure

In operation, the fixed and measured quantities determine part flow area employing two measurement sequences. Plenum volumetric airflows are initially reconciled for Pitot stagnation pressures $P_o - P_o$ obtaining the plenums ratio of internal airflow velocities V_{P_1}/V_{r_1}. The quantities are then arranged into a ratio of plenum volumetric airflows that combined with gauge and differential pressure measurements P_{r_1}, P_{atm}, and $P_{P_1} - P_{r_1}$ permit expression of air-density-independent part flow area A_{P_2} of equation (8-2). Equation (8-3) describes sequential multisensor error propagation determined from the influence of analytical process equations (8-1) and (8-2) with the aid of Table 8-1. Part flow area error is accordingly the algorithmic propagation of four independent pressure sensor instrumentation errors in this two-sequence measurement example, where individual sequence errors are summed because of the absence of correlation between the measurements each sequence contributes to the part flow area determination.

$$\Delta P_o = (P_{P_1} + \tfrac{1}{2}\rho V_{P_1}^2) - (P_{r_1} + \tfrac{1}{2}\rho V_{r_1}^2) \qquad P_o \text{ equilibrium sequence} \qquad (8\text{-}1)$$

$$A_{P_2} = A_{P_1} \cdot \left[\frac{\rho - 2(P_{P_1} - P_{r_1})/V_{r_1}^2}{\rho + 2(P_{r_1} - P_{atm})/V_{r_1}^2} \right]^{1/2} \qquad \text{part flow area sequence} \qquad (8\text{-}2)$$

In the first sequence, an equalized Pitot pressure measurement ΔP_o is acquired defining Bernoulli's equation (8-1). The algorithmic influence of this pressure measurement is represented by the sum of its static mean plus single RSS error contribution in the first sequence of equation (8-3). The second measurement sequence is defined by equation (8-2), whose algorithmic error propagation is obtained from arithmetic operations on measurements P_{r_1}, P_{atm}, and $P_{p_1} - P_{r_1}$ represented as the sum of their static mean plus RSS error contributions in equation (8-3).

$$\varepsilon_{\Delta P_O} + \varepsilon_{AP_2} = \{\overline{\varepsilon_{mean\ \Delta P_O}}\ \%FS + \varepsilon_{\Delta P_O}\ \%FS\ 1\sigma\}_{1st\ sequence} \quad \text{error propagation} \quad (8\text{-}3)$$

$$+ \{|\tfrac{1}{2}|[\overline{\varepsilon_{mean\ \Delta P_{p1-r1}}} + \overline{\varepsilon_{mean\ P_{r1}}} + \overline{\varepsilon_{mean\ P_{atm}}}]\ \%FS$$

$$+ |\tfrac{1}{2}|[\varepsilon_{\Delta P_{p1-r1}}^2 + \varepsilon_{P_{r1}}^2 + \varepsilon_{P_{atm}}^2]^{1/2}\ \%FS\ 1\sigma\}_{2nd\ sequence}$$

$$= \{\overline{0.1\%FS} + 0.1\%FS\ 1\sigma\}_{1st\ sequence}$$

$$+ \{|\tfrac{1}{2}|[\overline{0.1} + \overline{0.1} + \overline{0.1}]\%FS$$

$$+ |\tfrac{1}{2}|[0.1^2 + 0.1^2 + 0.1^2]^{1/2}\ \%FS\ 1\sigma\}_{2nd\ sequence}$$

$$= \overline{0.25\%FS} + 0.186\%FS\ 1\sigma \qquad \text{8-bit accuracy}$$

For the first sequence of equation (8-3) only the differential Pitot stagnation pressure measurement $P_o - P_o$ is propagated as algorithmic error. In the following second sequence, part plenum inlet area A_{p_1}, air density ρ and reference plenum inlet velocity V_{r_1} all are constants that do not appear as propagated error. However, the square root exponent influences the mean and RSS error of the three pressure measurements included in equation (8-2) by the absolute value shown. Four nine-bit accuracy pressure measurements are accordingly combined by these equations to realize an eight-bit accuracy part flow area.

Figure 8-3 abbreviates the signal conditioning and data conversion subsystems developed in the previous chapters for the sequential architecture of this section, employing Setra capacitive pressure sensors, and the homogeneous sensor architecture of the following section using Yellow Springs Instruments RTD sensors. Although each of these examples are coincidentally implemented with sensors of the same type, mixed sensors in either would provide no alteration in error propagation.

8-3   HOMOGENEOUS MULTISENSOR ARCHITECTURE

Figure 8-4 illustrates an 80 inch hot-strip rolling mill for processing heated slabs of steel into coils of various gauge strip, where conservation of mass, momentum, and energy require strip velocity increases with gauge reduction at each consecutive stand F1 through F6. An important process performance indicator related to coil production is the thermal losses dissipated by up to 40,000 horsepower available from the electric machines. For example, performance is degraded for slabs entering the mill cooler than an optimum temperature, because any slab energy shortfall must be made up by greater than nominal electromechanical machine output with corresponding I^2R thermal losses. In practice, these losses

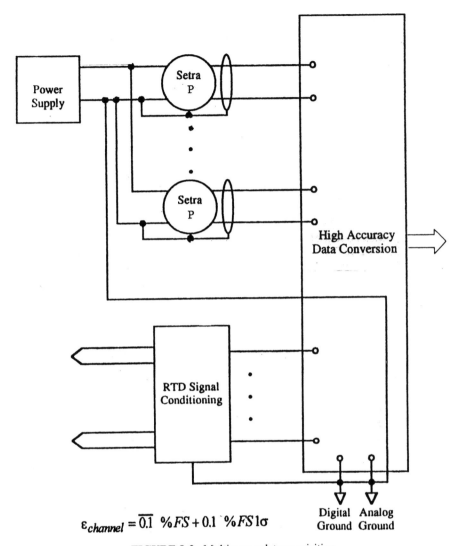

$$\varepsilon_{channel} = \overline{0.1}\ \%FS + 0.1\ \%FS\, 1\sigma$$

FIGURE 8-3. Multisensor data acquisition.

can total 1 megawatt for typical machine efficiencies of 97%, with 4,000,000 BTUs of heat requiring nonproductive mill standstill time for transfer to the environment.

Electric machine heating and cooling is usefully employed to predict required mill standstill time between coils to prevent machine temperatures from exceeding a safe target value above ambient. Pacing a mill for maximum production will accordingly be achieved at an optimum entering slab temperature for each steel hardness grade that minimizes standstill time. Relationships defining the $t_{standstill}$ quanti-

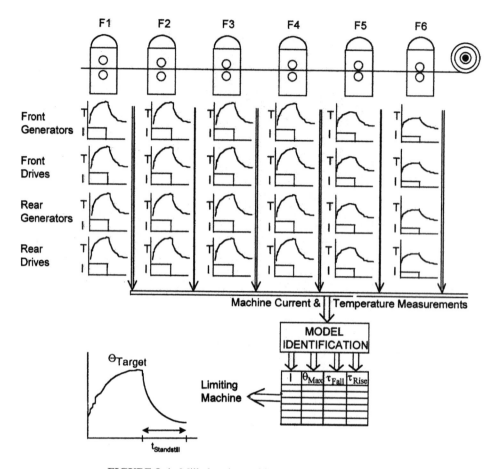

FIGURE 8-4. Mill electric machine temperature modeling.

ties are expressed by analytical algorithm equations (8-4) through (8-7) and Figure 8-5. Independent influences are observed for machine heating and cooling. The heating time constant for a machine is described by equation (8-4) as the ratio of its temperature rise time interval and its initial to rising difference in measured temperature slopes. The cooling time constant is shown by equation (8-5) from rearranging the temperature fall expression

$$\theta_{standstill} = (\theta_{target} - \theta_{ambient}) \cdot e[-(t_{standstill} - t_{standstill\ start})/\tau_{fall}] + \theta_{ambient}$$

The maximum steady-state machine temperature rise for continuous load application is predicted by equation (8-6). Table 8-3 further provides a thermal symbol glossary for these equations. Of primary interest is accounting for the algorithmic propagation of measurement errors in this homogeneous multisensor integration ex-

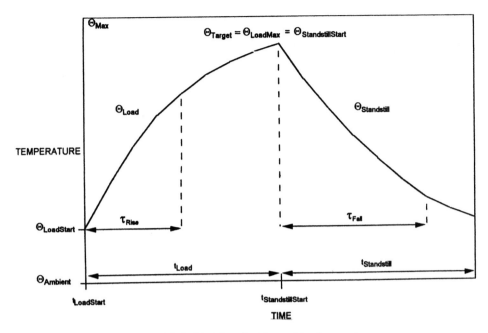

FIGURE 8-5. Limiting electric machine temperature.

ample from different equations whose error stackup is evaluated. Multiple electric machine temperature measurements are shown in Figure 8-4, each possessing a $\overline{0.1\%FS + 0.1\%FS}$ 1σ per channel instrumentation error from Figure 8-3, with algorithmic error propagation evaluated for the single highest temperature limiting machine illustrated by Figure 8-5. Note that θ_{target} temperature values appearing in analytical algorithm equations (8-5) and (8-7) of this example are constants, and therefore omitted from their corresponding error propagation equations (8-9) and (8-11). Only measurements can contribute error values.

TABLE 8-3. Electric Machine Thermal Glossary

Symbol	Comment
θ_{max}	Machine heating temperature prediction at $t = \infty$
θ_{target}	Defined machine temperature limit constant
$\theta_{load\ max},\ \theta_{standstill\ start}$	Measured machine temperature at end of heating
$\theta_{load},\ \theta_{load\ start},\ \theta_{standstill}$	Measured running machine temperature
$\theta_{ambient}$	Measured machine inlet air temperature
τ_{rise}	Machine heating time constant
τ_{fall}	Machine cooling time constant
$\tau_{standstill}$	Machine cooling interval prediction

Analytical algorithm equations:

$$\tau_{\text{rise}} = \frac{t_{\text{load}} - t_{\text{load start}}}{\ln\left[\frac{d}{dt}\theta_{\text{load}}\Big|_{t_{\text{load}}=t_{\text{load start}}}\right] - \ln\left[\frac{d}{dt}\theta_{\text{load}}\Big|_{t_{\text{load}}\neq t_{\text{load start}}}\right]} \tag{8-4}$$

$$\tau_{\text{fall}} = \frac{-(t_{\text{standstill}} - t_{\text{standstill start}})}{\ln\left(\dfrac{\theta_{\text{standstill}} - \theta_{\text{ambient}}}{\theta_{\text{target}} - \theta_{\text{ambient}}}\right)} \tag{8-5}$$

$$\theta_{\text{max}} = \left[\tau_{\text{rise}} \cdot \left(\frac{d}{dt}\theta_{\text{load}}\Big|_{t_{\text{load}}=t_{\text{load start}}}\right)\right] + \theta_{\text{load start}} \tag{8-6}$$

$$t_{\text{standstill}} = \left\{(-\tau_{\text{fall}}) \cdot \left[\ln\left(\frac{(\theta_{\text{target}} - \theta_{\text{max}}) \cdot e^{\left(\frac{t_{\text{load}}-t_{\text{load start}}}{\tau_{\text{rise}}}\right)} + (\theta_{\text{max}} - \theta_{\text{ambient}})}{(\theta_{\text{target}} - \theta_{\text{ambient}})}\right)\right]\right\}$$
$$+ t_{\text{standstill start}} \tag{8-7}$$

Error propagation equations:

$$\varepsilon_{\tau_{\text{rise}}} = \left\{\overline{\varepsilon_{\text{mean }\theta_{\text{load start}}}}\%FS + \varepsilon_{\theta_{\text{load start}}}\%FS1\sigma\right\}_{1\text{st sequence}} \tag{8-8}$$
$$+ \left\{\overline{\varepsilon_{\text{mean }\theta_{\text{load}}}}\%FS + \varepsilon_{\theta_{\text{load}}}\%FS1\sigma\right\}_{2\text{nd sequence}}$$
$$= \overline{0.2}\%FS + 0.2\%FS1\sigma$$

$$\varepsilon_{\tau_{\text{fall}}} = \left[\overline{\varepsilon_{\text{mean }\theta_{\text{standstill}}}} + 2\overline{\varepsilon_{\text{mean }\theta_{\text{ambient}}}}\right]\%FS \tag{8-9}$$
$$+ \left[\varepsilon^2_{\theta_{\text{standstill}}} + 2\varepsilon^2_{\theta_{\text{ambient}}}\right]^{1/2}\%FS1\sigma$$
$$= \overline{0.3}\%FS + 0.17\%FS1\sigma$$

$$\varepsilon_{\theta_{\text{max}}} = \left[\overline{\varepsilon_{\text{mean }\tau_{\text{rise}}}} + 2\overline{\varepsilon_{\text{mean }\theta_{\text{load start}}}}\right]\%FS \tag{8-10}$$
$$+ \left[\varepsilon^2_{\tau_{\text{rise}}} + 2\varepsilon^2_{\theta_{\text{load start}}}\right]^{1/2}\%FS\sigma$$
$$= \overline{0.4}\%FS + 0.17\%FS1\sigma$$

$$\varepsilon_{t_{\text{standstill}}} = \left[\overline{\varepsilon_{\text{mean }\tau_{\text{fall}}}} + 2\overline{\varepsilon_{\text{mean }\theta_{\text{max}}}}\right. \tag{8-11}$$
$$+ \left|\overline{\varepsilon_{\text{mean }\tau_{\text{rise}}}}\right| + 2\overline{\varepsilon_{\text{mean }\theta_{\text{ambient}}}}\right]\%FS$$
$$+ \left[\varepsilon^2_{\tau_{\text{fall}}} + 2(\varepsilon_{\theta_{\text{max}}})^2 + |\varepsilon_{\tau_{\text{rise}}}|^2 + 2\varepsilon^2_{\theta_{\text{ambient}}}\right]^{1/2}\%FS1\sigma$$
$$= \overline{1.50}\%FS + 0.38\%FS1\sigma \qquad \text{6-bit accuracy}$$

Mapping equation (8-4) to (8-8), observing Table 8-1, involves two temperature measurements for the conditions load start and load at different times, denoted by the first and second sequences in evaluating the limiting machine heating rise time-constant error. Mapping equation (8-5) to (8-9) involves summing one machine temperature measurement error at standstill with two ambient temperature entries for the machine cooling fall time–constant error evaluation. Mapping equation (8-6) to (8-10) requires summing the previous rise time–constant error plus two load start temperature error entries to define the error of the maximum predicted machine temperature. Standstill analytical algorithm and error propagation equations (8-7) and (8-11) combine the foregoing evaluations in four entries, including the rise time–constant within the exponent that is treated as a multiplicand and summed by Table 8-1. Ancillary mathematical operations in equations (8-4) through (8-7), including ln functions of arguments, accordingly have no influence on error propagation. Total measurement error equivalent to six-bit accuracy is dominated by the aggregation of repetitively propagated mean error values revealing their pronounced influence.

8-4   HETEROGENEOUS MULTISENSOR ARCHITECTURE

Challenges to contemporary process control include realizing the potential of in situ sensors and actuators applied beyond apparatus boundaries to accommodate increasingly complex process operations. The relationship between process and control design generally involves process design for controllability, with stability provided by the control compensator design. Uniform processing effectiveness requires attenuating variability, disorder, and disturbances, which is aided by process decomposition into a natural hierarchy of linear and decoupled influences that link environmental, in situ, and product subprocesses. It is significant that control performance for a system cannot achieve less variability than the uncertainty expressed by its total instrumentation error regardless of control sophistication. Real-time process measurements offer both model updating and minimization of processing disorder through feedback regulation. Further, accurate process models enable useful feedforward control references for achieving reduced disturbance state progression throughout a processing cycle.

Pulsed laser deposition (PLD) is a versatile thin-film manufacturing process for applications ranging from MoS_2 space tribological coatings to YCBO high-T_c superconductor buses whose modular process control implementation is illustrated by Figure 8-6. High-power excimer laser-ablated target material generates an intermediate plume subprocess of ions and neutrals for substrate deposition within a high vacuum chamber whose dynamics are only partially understood with regard to film growth. This example system employs feedback control of laser energy density e and repetition rate p based upon in situ microbalance-sensed deposition thickness m and spectroscopic plume density a. These relationships are described by equations (8-12) to (8-14). A hierarchically defined PLD subprocess control structure is shown in Figure 8-7 whereby energy transformations dominate the environmental

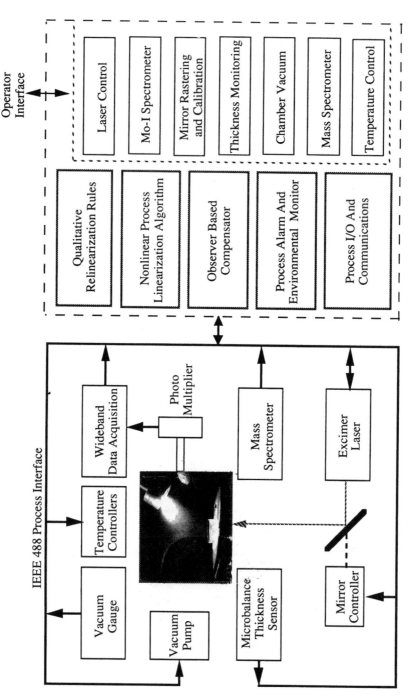

FIGURE 8-6. Modular pulsed laser deposition system.

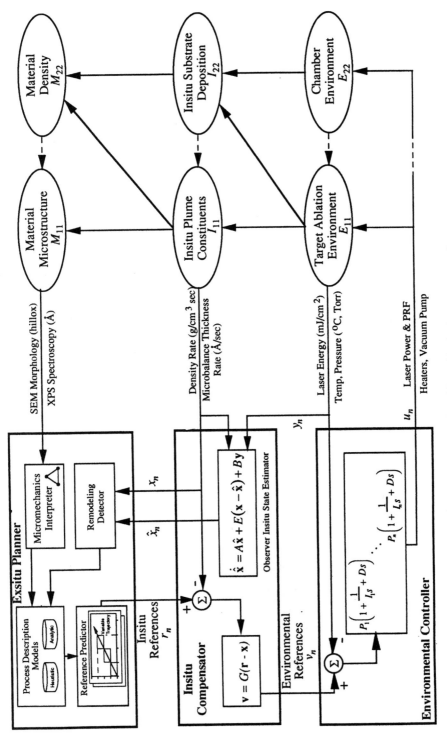

FIGURE 8-7. PLD hierarchical subprocess control.

to in situ subprocess influence mapping, and material properties the in situ to product subprocess mapping.

$$\dot{m} = \Sigma k_m \exp\left\{-\frac{|m - \mu_{mh}|^2}{2\sigma_{mh}^2} - \frac{|a - \mu_{ah}|^2}{2\sigma_{ah}^2} - \frac{|e - \mu_{eh}|^2}{2\sigma_{eh}^2} - \frac{|p - \mu_{ph}|^2}{2\sigma_{ph}^2}\right\} \qquad (8\text{-}12)$$

$$\dot{a} = \Sigma k_a \exp\left\{-\frac{|m - \mu_{mh}|^2}{2\sigma_{mh}^2} - \frac{|a - \mu_{ah}|^2}{2\sigma_{ah}^2} - \frac{|e - \mu_{eh}|^2}{2\sigma_{eh}^2} - \frac{|p - \mu_{ph}|^2}{2\sigma_{ph}^2}\right\} \qquad (8\text{-}13)$$

$$\begin{bmatrix} \dot{m} \\ \dot{a} \end{bmatrix} = \begin{bmatrix} f_1(m, a, e, p) \\ f_2(m, a, e, p) \end{bmatrix} = \frac{\partial f}{\partial(m, a)}\bigg|_{\substack{m_0, a_0 \\ e_0, p_0}} \begin{bmatrix} \Delta m \\ \Delta a \end{bmatrix} + \frac{\partial f}{\partial(e, p)}\bigg|_{\substack{m_0, a_0 \\ e_0, p_0}} \begin{bmatrix} \Delta e \\ \Delta p \end{bmatrix} \qquad (8\text{-}14)$$

where
m = microbalance sensed thickness (Å)
a = spectrometer sensed plume density (g/cc)
e = laser energy density (mJ/cm^2)
p = laser pulse repetition rate (Hz)

   Due to only marginal adequacy of PLD analytical process models, however, an alternative empirical model obtained from factorial process data is described in Figure 8-8. From that data set, a radial basis function fit of equations (8-12) and (8-13) provides the linearized differential equation approximation of equation (8-14). This control algorithm is assisted by an observer shown in Figure 8-7, whose state estimates $\hat{x}$ are compared with actual in situ sensor data x to detect when process migration is sufficient to require relinearization of the control algorithm. Plume density rate and deposition thickness rate data provide additional process knowledge useful for feedback control of film growth. Table 8-4 defines the decoupled subprocess influences of Figure 8-7 by their zero off-diagonal hierarchical mapping matrices, which substantially account for the effectiveness of the PLD deposition process. The merit of subprocess decoupling is in reduced iteration of controlled variables and required control complexity. Note that environ-

TABLE 8-4. PLD Subprocess Mapping

$$\begin{bmatrix} \text{SEM Morphohgy (hillox)} \\ \text{XPS Spectroscopy (Å)} \end{bmatrix} = \begin{bmatrix} M_{11} & 0 \\ M_{21} & M_{22} \end{bmatrix} \begin{bmatrix} \text{Density rate (g/cm}^3\text{ sec)} \\ \text{Microbalance thickness rate (Å/sec)} \end{bmatrix}$$

$$\begin{bmatrix} \text{Density rate (g/cm}^3\text{ sec)} \\ \text{Microbalance thickness rate (Å/sec)} \end{bmatrix} = \begin{bmatrix} I_{11} & 0 \\ I_{21} & I_{22} \end{bmatrix} \begin{bmatrix} \text{Laser energy, (mJ/cm}^2\text{)} \\ \text{Temp, pressure (°C, Torr)} \end{bmatrix}$$

$$\begin{bmatrix} \text{Laser energy, (mJ/cm}^2\text{)} \\ \text{Temp, pressure (°C, Torr)} \end{bmatrix} = \begin{bmatrix} E_{11} \\ 0 \end{bmatrix} \begin{bmatrix} \text{Laser power \& PRF} \\ \text{Heaters, vacuum pump} \end{bmatrix}$$

mental subprocess parameters exhibit the least coupling, and the final material parameters are evaluated ex situ offline by scanning electron microscopy (SEM) and X-ray photon spectroscopy (XPS).

Heterogeneous multisensor data permits the integration of nonoverlapping information from different sources, including nonredundant achievement of improved data characterization, and process feature identification unavailable from single sensors. Previous chapters have described instrumentation designs for sensors that in this example are characterized as environmental measurements, such as energy, temperature, and pressure. Sensor attribution is provided with in situ subprocess data acquired from a quartz crystal microbalance (QCM) and optical emission spectrometer (OES) beyond apparatus boundaries. An Inficon QCM measures film thickness online to 10 Angstroms by crystal frequency changes from deposited mass buildup based upon equation (8-15), with an error of approximately 3%FS verified by offline ex situ SEM measurement corresponding to five-bit accuracy from Table 6-2. Optical emission spectroscopy of the plume subprocess provides a real-time chemistry measurement alternative to mass spectroscopy, enabled by wideband digitization, for an improved process control capability. This measurement is shown in Figure 8-9, which shows chemical species

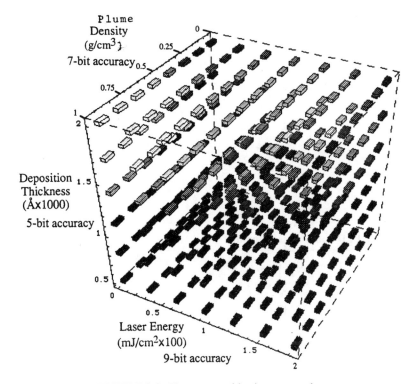

FIGURE 8-8. Hyperspectral in situ process data.

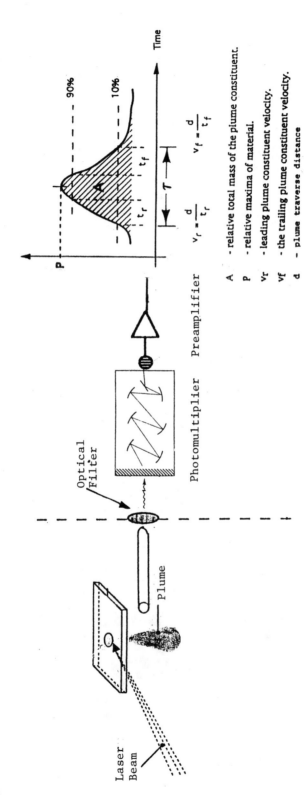

FIGURE 8-9. Plume optical emission spectrometer.

$$\frac{\text{Scope} f_s}{\text{Plume BW}} = \frac{\dfrac{400 \text{ MHz}}{2}}{1.25 \mu\sec_\tau} = 250 \frac{\text{Hz}}{\text{Hz}} \quad 7\text{-bit accuracy} \quad (6\text{-}13)$$

184

selected by specific filter elements. Employing a 400 megasample digital storage oscilloscope provides Nyquist sampling of the 200 MHz photomultiplier sensor, such that 1.25 microsecond width plume emissions, following nanosecond pulsed-laser target ablations, yield plume density waveform measurements of seven-bit accuracy by equation (6-13) for an f_s/BW ratio of 250 with reference to Tables 4-2 and 6-2.

$$t_f = \left[\frac{N_q d_q}{\pi d_f f_c C} \right] \qquad (8\text{-}15)$$

where
  t_f = film thickness (cm)
  d_q = quartz density (g/cm^3)
  N_q = crystal frequency constant (Hz/cm)
  d_f = film density (g/cm^3)
  f_c = coated crystal frequency (Hz)
  C = calibration constant (1/cm^2)

The hierarchical process control schema of Figure 8-7 additionally shows a system structured according to an increasing process knowledge representation at decreasing accuracy with subprocess ascension, and vice versa, analogous to Heisenberg's uncertainty principle. For example, in situ process measurements acquire higher information content energy and matter transformations such as the five-bit accuracy QCM thickness and seven-bit accuracy OES plume density sensors. These are in contrast to the limited information content of temperature and pressure environmental process measurements available to nine-bit accuracy from Figure 8-3. Regardless of the fact the five-bit QCM measurement accuracy dominates both the data model of Figure 8-8 and control algorithm of equation (8-14), there is no performance loss because of the higher attribution revealed in this hyperspectral spatial representation, including per-axis data accuracy, with the in situ data feature space providing system identification. The utility of system identification is in determining control operating values experimentally when analytical process models are inadequate. With the empirical data model of Figure 8-8, optimum process operation is featured in the upper left data region, where specific laser energy values are identified that beneficially maximize plume density and deposition thickness.

  This process control example also emphasizes the merit of system implementations employing the instrumentation hierarchy defined by Figure 1-18, where the realization of performance capabilities is enhanced by matching the signal attribution at each level. The immediacy of the corresponding signal models provide useful descriptive functions that increasingly are applied in a substitutive role, in place of describing processing specifications and incomplete process models, to enable the utilization of evolving complex process knowledge online for improved processing results.

BIBLIOGRAPHY

1. D. L. Hall, *Mathematical Techniques in Multisensor Data Fusion,* Norwood, MA: Artech House, 1992.
2. J. Llimas and E. Waltz, *Multisensor Data Fusion,* Norwood, MA: Artech House, 1990.
3. S. R. LeClair, "Sensor Fusion: The Application of Artificial Intelligence Technology to Process Control," *Proceedings FORTH Conference,* 1986.
4. R. L. Shell and E. L. Hall, *Handbook of Industrial Automation,* New York, Marcel Dekker, 2000.
5. P. H. Garrett, J. J. Jones, and S. R. LeClair, "Self-Directed Processing of Materials," *Elsevier Engr. Appl. Artificial Intelligence, 12,* 1999.
6. D. Bobrow, *Qualitative Reasoning About Physical Systems,* Cambridge, MA: MIT Press, 1985.
7. *Abstracts of Multisensor Integration Research,* NSF Workshop, Div. of Mfg., Snowbird, 1987.
8. B. K. Hill, "High Accuracy Airflow Measurement System," M.S. Thesis, Electrical Engineering, University of Cincinnati, 1990.
9. R. A. Oswald, "Finishing Mill Electric Machine Expert Advisor for Production Optimization," M.S. Thesis, Electrical Engineering, University of Cincinnati, 1995.
10. S. J. P. Laube, "Hierarchical Control of Pulsed Laser Deposition Processes for Manufacture," Ph.D. Dissertation, Electrical Engineering, University of Cincinnati, 1994.
11. T. C. Henderson, et al., "Multisensor Knowledge Systems," Technical Report, University of Utah, 1986.
12. *Abstracts of Manufacturing Systems Integration Research,* NSF Workshop NSF-G-DMC 8516526, St. Clair, November 1985.
13. H. F. Durrant-Whyte, "Sensor Models and Multisensor Integration," *Intl. J. Robotics Research, 6*(3): 3, 1987.

9

INSTRUMENTATION SYSTEM INTEGRATION AND INTERFACES

9-0 INTRODUCTION

Technical evolution and economic influences have combined to define the integration of contemporary multisensor instrumentation systems relative to a delineation of applications. A hierarchical instrumentation taxonomy is accordingly described as illustrated by discrete automatic test equipment, remote measurement environments, automation system virtual instruments, and analytical instrumentation for aiding sensed-feature understanding. The integration of each of these instrumentation categories is also defined by bus and network structures appropriate for meeting application performance requirements.

Chapter highlights include the description of virtual instrument capabilities for elevating fundamental sensor data to a higher attribution, enabling more complex cognitive interpretation. Such attribution is then extended to analytical instrumentation employing hyperspectral sensing of multiple spatial and spectral data for improved feature characterization. This is shown to be useful in advanced process control systems for comparing product states to goal states during manufacturing for the purpose of synthesizing compensating online quality control references.

9-1 SYSTEM INTEGRATION AND INTERFACE BUSES

Electrical measurement has been evolving for nearly two centuries since the invention of the galvanometer in 1820. Continued development has provided an expanding range of sophisticated measurement, signal conditioning, analysis, and data presentation capabilities with the instrumentation taxonomy, shown in Figure 9-1, that can accommodate the comprehensive data requirements of advanced hierarchical sensor and actuator systems. Four distinct instrumentation integration structures are defined, each of which involve different implementations for meeting their re-

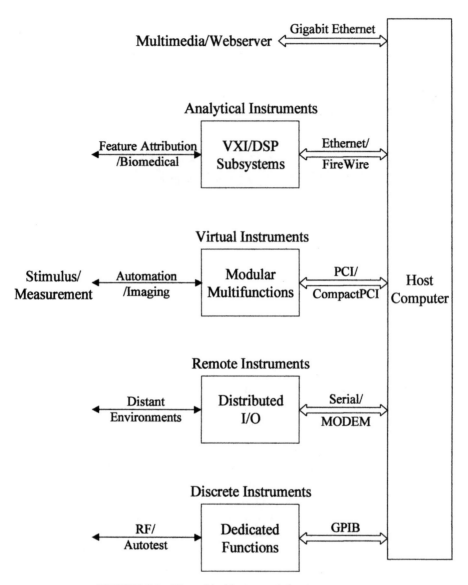

FIGURE 9-1. Hierarchical instrumentation taxonomony.

spective excitation and measurement applications. Examples are presented in the sections that follow that highlight effective solutions to contemporary instrumentation challenges for each of these architectures.

The diversity of existing bus structures provides a useful delineation of capabilities for instrumentation system integration. Figure 9-2 introduces basic computer bus classifications. Level-0 traces describe intercomponent board connections that

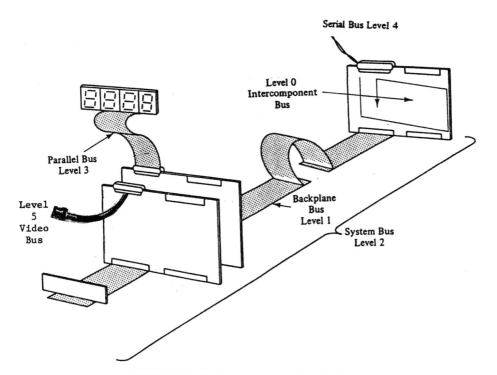

FIGURE 9-2. Basic computer bus classifications.

are characterized by signals specific to their digital devices. Level-1 dedicated bus-es, such as the industry standard architecture (ISA) bus, provide buffered subsystem peripheral component interfacing, including protocols to accommodate signal prop-agation delays. Level-2 system buses, such as the peripheral component intercon-nect (PCI) structure detailed in Figure 9-11, offer comprehensive bus master ser-vices, including arbitration and concurrent operation. Level-3 parallel buses enable peripheral extensions for Level-1 buses, including the general purpose interface bus (GPIB) and small computer systems interface (SCSI) bus. Level-4 serial buses are the longest structures in the bus repertoire, and range from early standards such as RS-232C to the more recent universal serial bus (USB) described in the following section. Serial bus transmission protocols are divided into synchronous and asyn-chronous modes, with the latter prevalent. The Level-5 video bus may be limited to an AGP port that supports the monitor.

The GPIB bus has achieved acceptance since its introduction by Hewlett-Packard because of its robustness for networking discrete instruments. This parallel bus can link 15 instruments plus a controller with 16 active lines, eight for data and eight for control, as shown in Figure 9-3. Communication control procedures initi-ated prior to data transmission designate transmitting instruments and receiving in-struments. Instead of address lines, there are three data-transfer and five bus man-

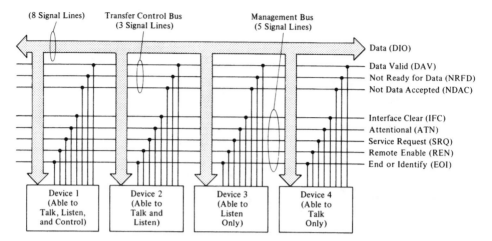

FIGURE 9-3. GPIB parallel bus structure.

agement lines for communication utilities. When ATN is high, all instruments must listen to the DIO lines. When ATN is low, only designated instruments can send and receive data.

External information exchanges with the host computer for all of the instrumentation architectures of Figure 9-1 can be aided by the Gigabit Ethernet, especially when high resolution graphics are involved. The efficiency of the Gigabit Ethernet relies upon full-duplex transmission employing all four wire pairs of common Category 5 cable, plus enabling terminal equipment shown in Figure 9-4. Performance is facilitated by five-level PAM coding, Trellis forward error correction, and DSP received signal equalization. Conventional Ethernet parameters are also introduced in the following section.

Computer-based automatic test equipment (ATE) has evolved as an effective application of parallel buses to link modular instruments in a systematic quality control structure for evaluating and documenting the performance of complex electronic systems, which may also include radio frequency signals. This structure is illustrated by the example of Figure 9-5 for discrete units under test, such as exercised during the preflight countdown of the Space Shuttle. Compared with manual stimulus and measurement, ATE offers improved test productivity, consistent test repetition with objective results, and more comprehensive test options and durations. Contemporary ATE software test executives typically are multisequence programs in both scripted and graphical languages, such as C++ and LabVIEW, with automatic report generation to ASCII, HTML, and database files including Access and SQL Server. The abbreviated test language for all systems (ATLAS) is an IEEE standard that was created for aviation electronic system maintenance, and eventually adapted to many ATE applications. ATE programs typically consist of macros with symbolic parameters that are combined by a linker to implement test applications.

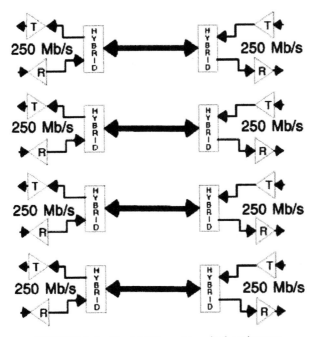

FIGURE 9-4. Gigabit Ethernet terminal equipment.

9-2 INSTRUMENT SERIAL BUS INTERFACES

Digital serial baseband signaling provides the majority of peripheral device and instrumentation system connections to host computers. Local area networks (LANs) have distinct functionalities, basically described by the network access devices that interface users to interconnecting media. For example, computer LANs integrate network access devices internally into hosts and servers such as universal asynchronous receiver and transmitter (UART) terminal devices. This structure is described by Figure 9-6. Source encoding commonly uses the RS-232C standard, shown as a full-duplex, null-MODEM connection by Figure 9-7, that is capable of data rates to 115 Kbps and distances to 50 feet. The speed versus distance for local area networks is principally determined by the intersymbol interference of adjacent bits, owing to the natural contraction of interconnecting media bandwidth with increasing distance. For noisy applications, RS-485 adds differential line drivers and receivers to UARTs, whose common mode interference rejection permits distances to 4000 feet, while supporting 32 active nodes per serial port. The higher performance universal serial bus (USB) offers low-cost consolidation of computer peripheral interfacing that can accommodate up to 127 peripheral devices with data rates to 12 Mbps. This is a polled bus utilizing packet data with automatic peripheral enumeration by its bus controller. However, USB hub-to-peripheral distances are limited to 15 feet. Closer source-encoded transmission usually is connected point-to-point,

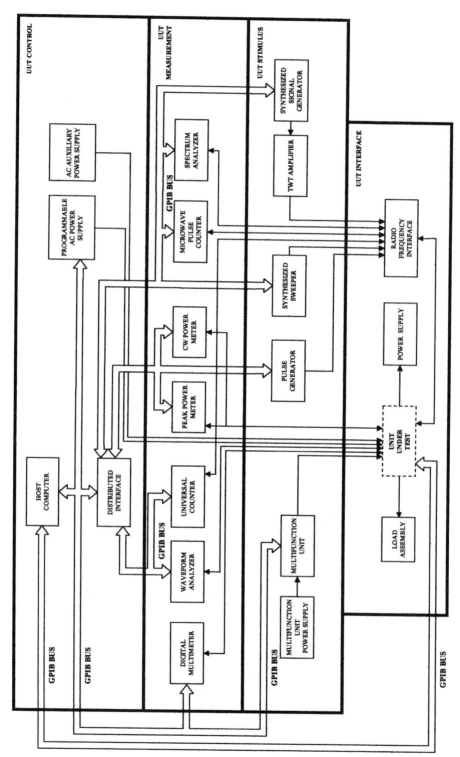

FIGURE 9-5. Discrete instrument parallel bus automatic test equipment.

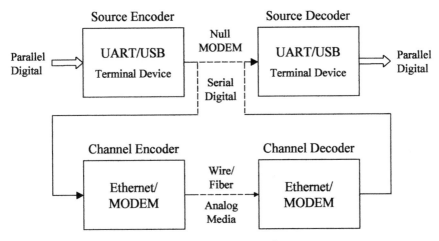

FIGURE 9-6. Serial bus network structure.

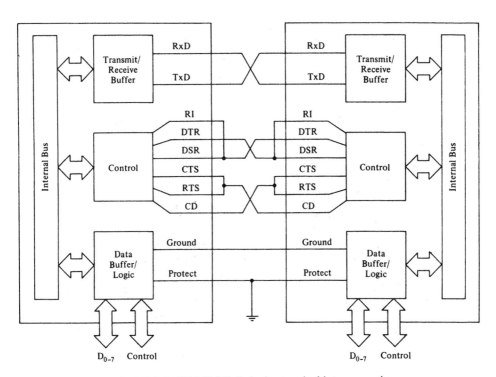

FIGURE 9-7. RS-232C Full-duplex terminal interconnection.

whereas extended channel-encoded transmission generally employs a multinode bus topology.

Alternatively, public LANs rely upon external network access devices such as Ethernet. Ethernet is a universal network currently employed worldwide because of advances in performance to 100 Mbps and, separately, economy of implementation enabled by twisted pair connectivity. This LAN further offers the versatility of coax, twisted pair, and fiber media. Its carrier-sense multiple access, collision detection (CSMA/CD) datalink protocol benefits from simplicity and effectiveness. Frequently applied twisted-pair Ethernet (10 Base T) supports data rates to 10 Mbps, whereas fast Ethernet employs fiber media (100 Base FX) supporting data rates to 100 Mbps. Ethernet employs a bus topology and packet data format with a 48-bit unique worldwide address and allowable message size ranging from 512 bits to 1512 bytes, where twisted-pair segments may extend to 1640 feet and fiber segments to 3600 feet. Note that Ethernet source encoding/decoding does not rely upon the terminal devices shown in Figure 9-6 because of its higher data rate. Gigabit Ethernet (1000 Base T4) utilizing four twisted pairs was described in the preceding section.

The growing number of process instrumentation and control systems from multiple vendors that require integration compatibility has led to the evolution of standardized public LANs for industrial applications that provide error checking and the economy of multinode device connectivity. These networks are exemplified by Foundation Fieldbus and the controller area network (CAN). Fieldbus employs twisted pair connectivity with a data rate of 31.25 Kbps and a transmission distance to 1 mile. It is intended for distributed process automation systems, and usefully permits remote devices to be powered over the same signal pair. CAN was initially designed to economically link onboard automotive digital functions. However, its low-speed and high-speed data rate options, respectively 125 Kbps and 1 Mbps, plus reliability provided by a multiple error checking protocol has resulted in a viable industrial network for distances to 1640 feet.

An emerging process instrumentation and control network concept is to permit system nodes to communicate directly without passing through a host computer as conventionally required. This autonomous capability redefines the host in a supervisory capacity, enabling network assets to be reallocated as process priorities require. Such a local operating network (LON) protocol is offered by Echelon Corporation as LonWorks and configured under LonMaker for Windows. LonWorks employs serial packet data exchange over twisted pairs in a bus topology at data rates of 78 Kbps to 4000 feet and 1.25 Mbps to 400 feet.

The transmission of digital data over media lengths greater than 1 mile requires additional complexity to overcome the distance limiting factors associated with intersymbol interference. The addition of a channel encoder modulator and demodulator (MODEM) provides a solution to this limitation by encoding serial baseband signals in a modulation format optimized for extended media. Commercial MODEMs are frequently interfaced by the RS-232C standard, and offer both synchronous and asynchronous bit-serial transmission. Modulation formats include frequency shift keyed (FSK) and quadrature phase shift keyed (QPSK). MODEM transmission errors are primarily a result of noise bursts, especially over wireless links, lasting from 1–50 milliseconds and occurring at random. Table 9-1 describes the ASCII character set frequently utilized in bit-serial data transmission.

The application of remote sensing instruments is diverse and ranges from hostile environments such as nuclear reactors to down-hole oil exploration to spacecraft to the electronic battlefield. The prevailing connectivity for this architecture, defined in Figure 9-1, employs serial data networks that meet specific data rate and distance requirements. A satellite radiometer remote instrument example is shown in Figure 9-8, including a serial bus interfaced telemetry MODEM. Total power millimeter wavelength radiometer spectrometers achieve a noise-equivalent temperature sensitivity (NEΔT) capable of sensing differences between surface temperatures, snow cover, moisture, and vegetation through clouds and dust where infrared sensors are not usable. Measured atmospheric noise power spectra acquired by this passive scanning instrument are heterodyned to centimeter wavelengths to facilitate amplification and filtering. The detection of amplified noise signals by square-law devices provide noise-equivalent temperatures with a beneficially high noise measurement sensitivity relative to internal receiver noise. By equations (9-1) and (9-2), the received noise power P in a defined receiver bandwidth $B(f, T)$, per solid angle Ω of antenna aperture $A(\theta, \phi)$, yields a radiometric temperature equivalence T. Noise-equivalent temperatures to 300°K are achievable with a 1° K measurement error.

TABLE 9-1. ASCII Character Set

		b_7	0	0	0	0	1	1	1	1
		b_6	0	0	1	1	0	0	1	1
Binary Code		b_5	0	1	0	1	0	1	0	1
b_4 b_3 b_2 b_1			Nonprintable		Printable Characters						
0	0	0	0	NUL	DLE	SPACE	0	@	P	\	p
0	0	0	1	SOH	DC1	!	1	A	Q	a	q
0	0	1	0	STX	DC2	//	2	B	R	b	r
0	0	1	1	ETX	DC3	#	3	C	S	c	s
0	1	0	0	EOT	DC4	$	4	D	T	d	t
0	1	0	1	ENQ	NAK	%	5	E	U	e	u
0	1	1	0	ACK	SYN	&	6	F	V	f	v
0	1	1	1	BEL	ETB	/	7	G	W	g	w
1	0	0	0	BS	CAN	(	8	H	X	h	x
1	0	0	1	HT	EM	)	9	1	Y	i	y
1	0	1	0	LF	SUB	*	:	J	Z	j	z
1	0	1	1	VT	ESC	+	;	K	[	k	{
1	1	0	0	FF	FS	'	<	L	\	I	:
1	1	0	1	CR	GS	–	=	M	]	m	]
1	1	1	0	SO	RS	.	>	N	^	n	~
1	1	1	1	SI	US	/	?	O	–	o	DEL

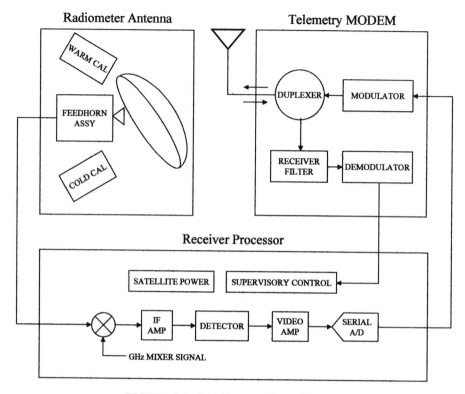

FIGURE 9-8. Serial bus satellite radiometer.

$$P = \frac{1}{2} \int_f \int_\Omega A(\theta, \phi)B(f, T)d\Omega \, df \qquad \text{watts} \qquad (9\text{-}1)$$

$$B(f, T) = \frac{2kTf^2}{c^2} \, \text{Hz} \qquad (9\text{-}2)$$

where
T = temperature, °K
k = Boltzmann constant, J/°k
c = velocity of light, m/s
f = noise frequency, Hz

A subsidiary performance issue for remote instruments is reliability of operation, especially when component replacement or maintenance are precluded. Reliability assessment provides an a priori calculated probability of continued operation for specified time intervals. This calculation is based upon component part experimental testing to acquire specific failure rate data, usually expressed as mean time be-

TABLE 9-2. Satellite Five Year Operation Prediction

Radiometer Element	MTBF (hrs)	$\exp\left(-\dfrac{44,000}{\text{MTBF}}\right)$ (Random Failure)
Antenna	9×10^5	0.952
Receiver	5×10^5	0.916
MODEM	7×10^5	0.939
Power	3×10^5	0.863
Reliability		$\Pi\ 0.707$

tween failures (MTBF) in hours, and applying this data to models of the physical behavior of complex systems of components over their life cycle. Three distinct intervals define components' end-to-end reliability life cycle: early failures attributable to component manufacturing defects; random failures arising from unmodeled disturbances over components useful life that are irreducible by replacement; and wearout failures occurring from components' performance depletion at their natural end of life.

For satellite reliability enhancement, early failures are routinely circumvented through accelerated component life-cycle testing and prelaunch replacement as necessary, in concert with the utilization of S-Grade electronic components, by applying established failure mode effects analysis (FMEA) procedures. Wearout is prevented by ensuring that total accumulated satellite operating time does not enter this life-cycle period for any of its components. Reliability is then predicted by the residual exponential behavior of component survival from random failures. Considering representative component MTBF values and a 44,000 hour operating period, described in Table 9-2, the total reliability for this series-connected system is determined by the product of the individual component reliabilities to reveal a 70% probability of operation over 5 years.

9-3   MICROWAVE MICROSCOPY VIRTUAL INSTRUMENT

The concept of computer-based instruments arose with the advent of inexpensive computation, furthered by the personal computer, that permitted networking discrete instruments into sophisticated automated test systems beginning in the 1970s. The evolution of more efficient data acquisition and presentation, resulting from user-defined programmability and reconfigurability, continues through the present to provide a more computationally intensive instrumentation framework. Contemporary virtual instruments are capable of elevating fundamental sensor data to a substantially higher attribution, enabling more complex cognitive interpretation. Multifunction I/O hardware is typically combined with application development software on a personal computer platform for the realization of specific virtual instruments like the following microwave microscopy example for sample assays in manufacturing and biomedical applications.

A benefit of microwave microscopy is micron resolution sample imaging of sub-surface as well as surface properties. With microwave excitation wavelengths on the order of millimeters, their limiting half-wavelength Abbe resolution barrier is extended through detection of shorter near-field evanescent microwave spatial wavelength components, aided by enhanced transducer excitation, sensitivity, and signal processing. With the virtual instrument of Figure 9-9, a 100 Hz sinusoidal IF signal dither of the 30 Ghz microwave source enables synthesis of 2F spatial frequency components associated with a sample. Sensitivity is increased by an automatic controller whose variable dc tunes the microwave source, as does the dither signal, to obtain a resonant frequency shift that maximizes the spatial components. The greater the 2F signal DFT-normalized magnitude achieved, the more selective the resonant curve quality factor and frequency shift images become relative to the sample. The resolution of this instrument is limited by typically –55dB of spurious noise from contemporary GHz sources, which is equivalent to nine-bit accuracy by Table 5-7.

The architecture of virtual instrument software may be divided into two layers: measurement and configuration services, and application development tools. Measurement and configuration services contain prescriptive software drivers for interfacing hardware I/O devices as subroutines that are usually accessed by graphical icons. Configuration utilities are also included in this layer for naming and setting hardware channel attributes such as amplitude scaling. Software selected for application development may be sourced separately from hardware devices only when compatibility is ensured. Examples of commercial virtual instrument software are listed in Table 9-3 for data acquisition, processing, presentation, and communications tasks. Graphical languages have become dominant for these systems owing to their speed of system prototyping, ease of data presentation, and self-documentation.

Graphical programs typically consist of an icon diagram, including a front panel that serves as the source code for an application program. The front panel provides a graphical user interface for functions to be executed concurrently. The LabVIEW diagram of Figure 9-10 shows a View Image module for generating the data display images shown in Figure 9-9,which also may be exported as Matlab files. When this program is initiated, the front panel (1) defines display visibility attributes. Assets within a "while loop" are then executed cyclically until control Done is set false (2), allowing conditional expressions to break this "while loop." The metronome icon describes a 50 millisecond interval within which the "while loop" iterates. The data structure that performs this image generation executes sequentially.

The concentric window (3) is analogous to a "case" statement in the C language, controlled by the Boolean Load variable. The "case" procedure occurs when Load is true. The icons within (3) allow a user to locate an image file, whereas the icons within (4) provide a subprogram for extracting microscopy image content for arrays representing frequency shift and quality factor data. A "sequence" data structure performs total image formatting. Icons within (4) select strings contained in (5) to write data images as XY vectors. Note that this code is set to display X and Y in millimeter units

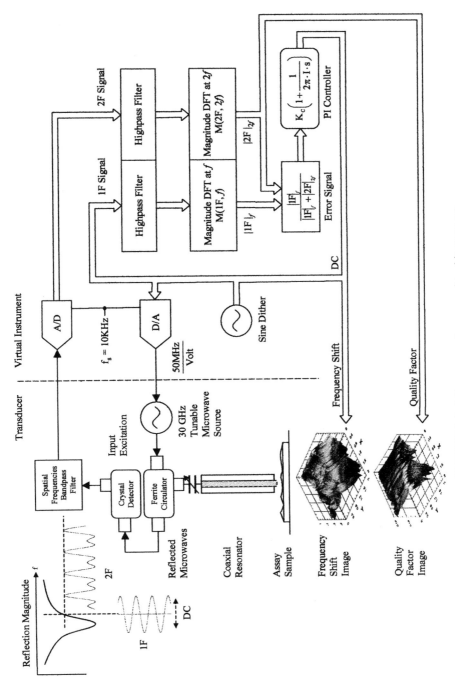

FIGURE 9-9. Microwave microsopy virtual instrument.

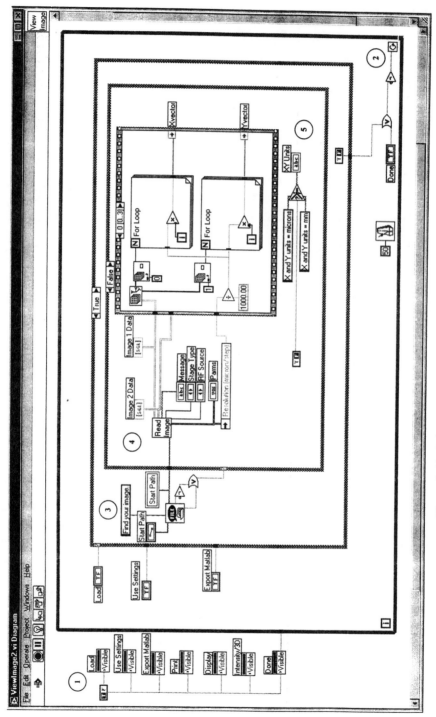

Figure 9-10. LabVIEW display generation graphical program.

TABLE 9-3. Example Instrumentation Graphical Software

Labtech Notebook (Labtech)
LabVIEW (National Instruments)
Agilant VEE (Agilant Technologies)
DASYLab (DSP Development Corp)
Snap-Master (Snapmaster)
	Graphical environments for
	  data acquisition and processing
	  development

DADiSP (DSP Development Corp) Visual C++ (Microsoft)	Graphical environments for data presentation development

The peripheral component interconnect is a versatile processor-independent computer bus structure, illustrated in Figure 9-11, that was introduced by Intel Corporation to enable CPU, memory, and peripheral device interconnections for peer-to-peer transfers of 64-bit words at up to 66 MHz rates, or 4 Gigabits per second, using burst packet transfers. None of the bus devices have dedicated memory or address assignments, but instead are configured and so assigned by BIOS flash memory on power-up. Power conserving reflected-wave logic switching is employed that requires only one-half logic level voltage excitation without the requirement for bus line impedance termination, but bus lengths must be short. Bus bridge extenders are accordingly used between separate PCI bus segments, and to other buses such as ISA, which permit concurrent separate bus operations. Up to 256 PCI buses can be supported with bridges, each with a maximum of 256 peripheral devices. CompactPCl is an industrially hardened modular PCI bus available in a 3U or 6U Eurocard form factor intended for embedded applications where robustness is essential. Implementations include communications servers, industrial automation, and defense electronic systems. The PCI bus typical bandwidth of 132 Mbytes per second supports video image manipulation, whereas the ISA bus bandwidth of 8 Mbytes per second cannot.

9-4   ANALYTICAL INSTRUMENTATION IN ADVANCED CONTROL

Computational instrumentation is described for real-time data applications with multisensor information systems featuring analytical ex situ planners applied to process control. Planners provide control advancement by assessing evolving measurements during processing to implement a global process real-time quality control loop. Ex situ planners are illustrated in Figures 8-7 and 9-12 that combine processing state evaluations with the synthesis of compensating control references for achieving product objectives, which in the aggregate represent inverse models of their respective processes. The example of Figure 9-12 shows a galvanizing line chromate coating system employing a scanning infrared reflectance, liquid film, in situ sensor for acquiring measurements beyond process apparatus boundaries. This

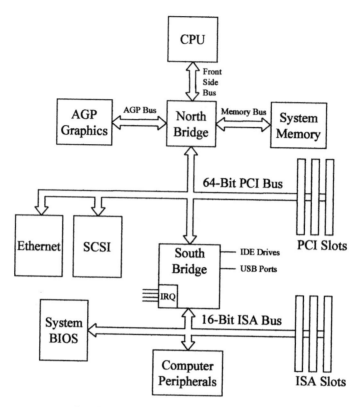

FIGURE 9-11. Exemplar PCI bus structure.

planner manages process adaptation by providing compensatory references to maintain processing goals.

During operation, a product chromate coat-weight goal in milligrams per square meter of strip area is achieved by measurement and control of a solute film thickness in milliliters per square meter. This is necessary to achieve galvanizing protection from white rust while preventing discoloration arising from excessive coat weight. Operational process disturbances introduced by squeegee force variations and strip speed changes are regulated, respectively, through conventional squeegee pressure and solute flow feedback controllers. Slower residual disturbances, such as introduced by squeegee wear or process parameter migration, can limit achievable coating quality that can be compensated for with feedforward references defining multiple rule-based planner episodes for restoring process equilibrium.

Analytical instrumentation for feature attribution can include more complex multidimensional microstructure assessments incorporating process parameter pattern recognition comparisons with processing goals. For example, thin-film deposition processes require increased data fenestration to describe crystalline growth mechanisms (bonds, evaporation, adsorption), physical properties (mass, phase,

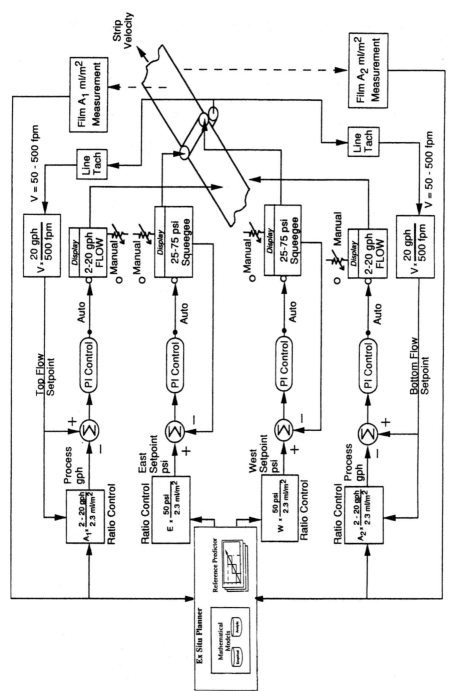

FIGURE 9-12. Ex situ planner coating disturbance compensator.

species), and structure (boundaries, geometry, morphology) in order to accurately define process actuation parameters (gas/liquid feedstock flows, heat, pressure). Implementation may be met as illustrated in Figure 9-13, employing a DSP accelerator and open-standard embedded VXI computers. VXI instrumentation back planes permit interoperability between different vendor hardware and a mix of software systems in order to combine technologies ranging from discrete instruments to multifunctional virtual instruments.

Following data acquisition and virtual instrument feature identification, a crystal facsimile classification is performed by means of an autoassociative network de-

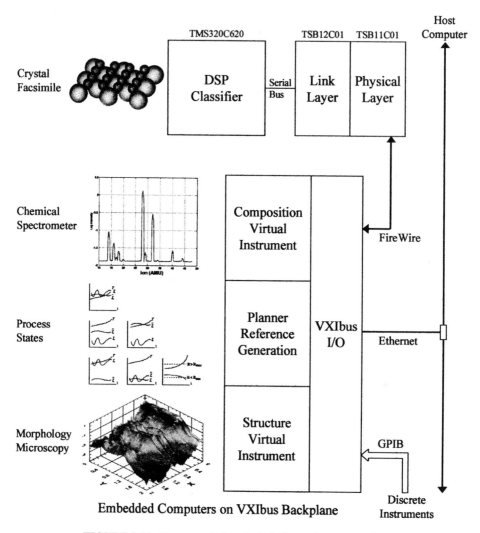

FIGURE 9-13. Hyperspectral analytical planner instrumentation.

scribed in Figure 9-14. This network is trained on crystalline prototypes to enable identification through crystal regularities. Autoassociative networks can be likened to content-addressable memory that can recognize correct output patterns given incomplete input patterns, and benefit from closed separation surfaces, unlike multilayer perceptron networks, which have open separation surfaces. Thus, autoassociative networks establish closed decision boundaries in the input space that capture the probability distributions of training crystal prototypes.

Real-time planner systems ordinarily operate at a higher level of abstraction than the interfaced control algorithms that complete a processing system. Further, hyperspectral imaging is a collateral methodology that integrates both spatially and spectrally continuous data to assist product characterization. The classification of a thin-film crystalline facsimile from multisensor chemical composition and morphology structure virtual features is expedited by DSP algorithms executing rapidly repeating sum of products operations. This is described by equation (9-3) and its associated code.

$$y_n = \sum_{i=0}^{N} A_{n-i} x_{n-i} \qquad (9\text{-}3)$$

where

y_n represents the output at time n

A_{n-i} represents a time-ordered set of static coefficients and

x_{n-i} represents the time-ordered set inputs, where i indicates an offset from time n

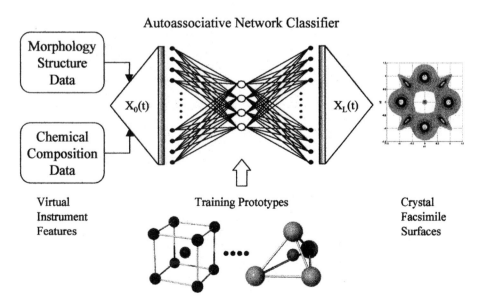

FIGURE 9-14. Autoassociative crystal feature classification.

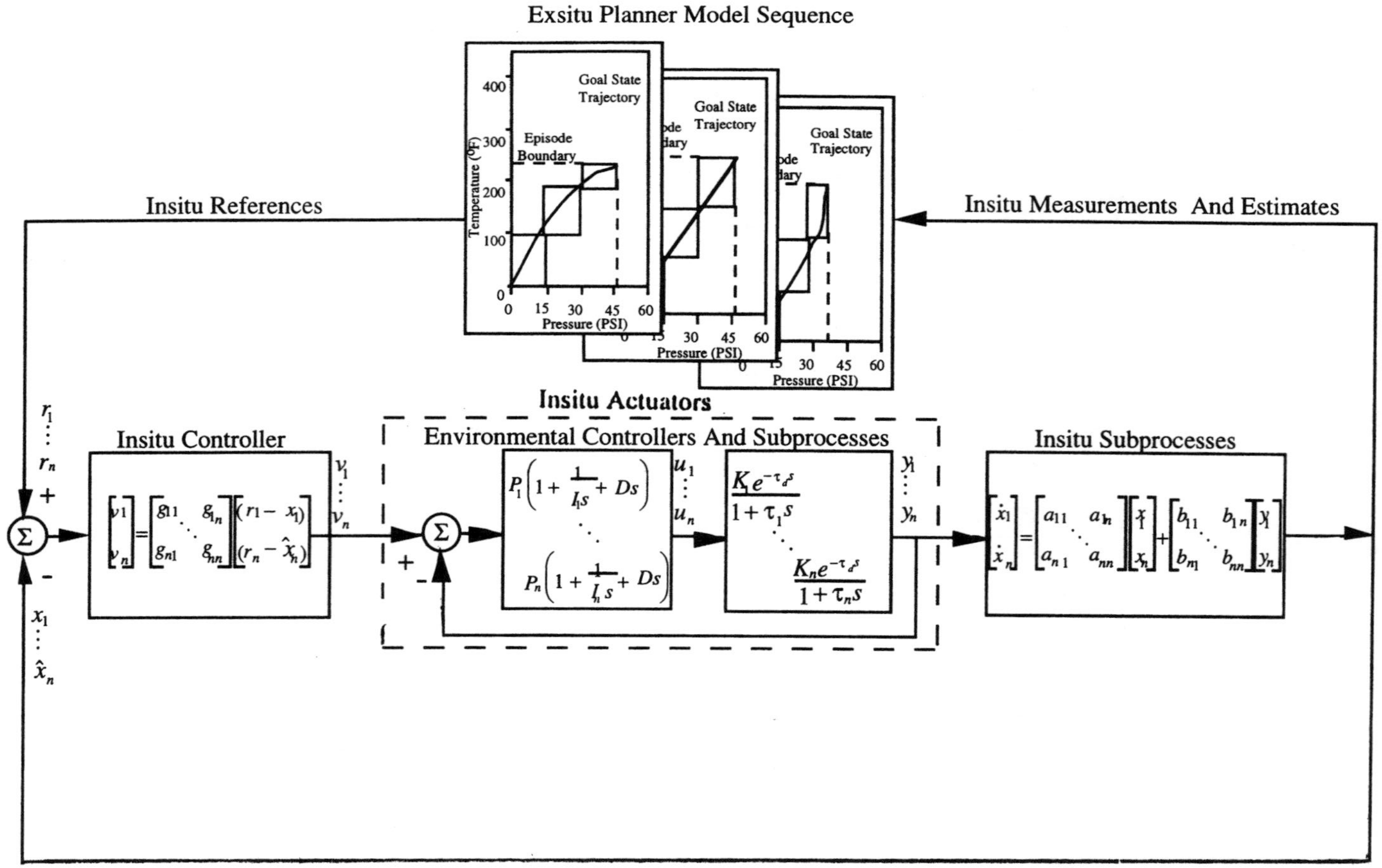

FIGURE 9-15. Planner-directed in situ control.

```
;  W = 32-bit memory words
;  C = Instruction cycles
;                          W C
 STM #x,AR2            ;  2 2
 STM #a,AR3            ;  2 2
 RPTZ A,#N-1           ;  1 3
 MAC *AR2+,  *AR3+,A  ;  1 1
 STH A,@y              ;  1 1
```

Ex situ planner outputs are synthesized by comparing online crystal classifications with a goal crystal prototype to provide compensating input references for the in situ controller shown in Figure 9-15. In situ state variables x_n are actuated by environmental subprocess transfer function excitation y_n in response to environmental references v_n that achieve a reduction in environmental time constants by the in situ controller gain magnitude, $\tau_n/(1 + g_{nn})$, owing to the environmental–in situ subprocess cascade relationship. This transformation provides extended in situ actuator bandwidths, enabling enhanced response of in situ variable $\dot{x}_n$ energy–mass–momentum state transitions, which is a principal contribution to the effectiveness of in situ control.

VXI embedded computer back plane I/O is internally served by a VXI bus with a GPIB interface for discrete instrument sensors. A local Ethernet link is provided for serving the host computer because of its 10/100 Mbps data rate and up to 1518 byte message capability. Ethernet employs a CSMA/CD network access algorithm that has negligible time delay or blocking when traffic is restricted by limiting connected nodes. Efficient wideband peer-to-peer digital interfacing up to 400 Mbps is available with an IEEE 1394 FireWire serial bus. FireWire supports both asynchronous and isochronous data transfer by means of a six-wire cable. Its physical layer tree topology is automatically reconciled with network node changes without host computer intervention, but interconnection distances are presently limited to 5 meters. FireWire data transfers are memory-based rather than channel-addressed to enable efficient processor-to-memory CPU transactions.

BIBLIOGRAPHY

1. T. L. Dean and M. P. Wellman, *Planning and Control,* San Mateo, CA: Morgan Kaufmann, 1991.
2. K. Forbus, et al., "Qualitative Spatial Reasoning: The Clock Project," *Artificial Intelligence, 51,* 1–3, 1991.
3. K. Osaki (Ed.), *JST Crystal Data,* Japan Science and Technology Corporation, 5-3 Yonbancho (Tokyo), 1997.
4. *Peripheral Design Handbook,* Santa Clara, CA: Intel Corporation, 1999.
5. W. Stallings, *Data and Computer Communications,* 3rd ed., New York: Macmillan, 1991.

6. R. F. Matejka, "Qualitative Process Automation Language," *Proceeding Aerospace Applications of Artificial Intelligence,* Air Force Research Laboratories, Dayton, Oct. 1989.

7. Y. H. Pao, *Adaptive Pattern Recognition and Neural Networks,* Reading, MA: Addison-Wesley, 1989.

8. W. J. Fabrycky and B. S. Blanchard, *Life-Cycle Cost and Economic Analysis,* Englewood Cliffs, NJ: Prentice Hall, 1991.

9. P. A. Laplante, *Real-Time Systems Design and Analysis,* Piscataway, NJ: IEEE Press, 1992.

10. P. Bhartia and I. J. Bahl, *Millimeter Wave Engineering and Applications,* New York: Wiley, 1984.

11. W. J. Wilson et al., "Millimeter Wave Imaging Sensor," *IEEE MTT-S Annual Digest,* 1986.

12. M. Bianchini et al., "Learning in Multilayered Networks Used as Autoassociators," *IEEE Transactions on Neural Networks, 6,* 512–515, March 1995.

13. R. C. Harney, "Practical Issues in Multisensor Target Recognition," *SPIE, Vol. 1300, Sensor Fusion III Conference,* 1990.

14. R. C. Gonzalez and R. E. Woods, *Digital Image Processing,* Reading, MA: Addison-Wesley, 1992.

15. S. K. Sin and C. H. Chen, "A Comparison of Deconvolution Techniques for the Ultrasonic Nondestructive Evaluation of Materials," *IEEE Transactions on Image Processing, 1,* 3–10, 1992.

INDEX